T0135720

Modeling and Analysis for Optimal Scheduling of Biodiesel Batch-Plants

Dissertation

zur Erlangung des akademischen Grades
Doktor-Ingenieur (Dr.-Ing.)

Genehmigt durch das
Zentrum für Ingenieurwissenschaften
der Martin-Luther-Universität Halle-Wittenberg
als organisatorische Grundeinheit für Forschung und Lehre im Range einer Fakultät

von
Herrn Mohamad Fauzan Amir, M. Eng.
geboren am 29.03.1977 in Kediri (Indonesien)

Geschäftsführender Direktor (Dekan): Prof. Dr.-Ing. Dr. h.c. Joachim Ulrich

Gutachter:
1. Prof. Dr.-Ing. Hans-Michael Hanisch
2. Prof. Dr. Zbigniew Banaszak

Halle (Saale), 14. Juli 2011

Bibliografische Information der Deutschen Nationalbibliothek

Die Deutsche Nationalbibliothek verzeichnet diese Publikation in der
Deutschen Nationalbibliografie; detaillierte bibliografische Daten sind
im Internet über http://dnb.d-nb.de abrufbar.

ISBN 978-3-8325-2933-8

Logos Verlag Berlin GmbH
Comeniushof, Gubener Str. 47,
10243 Berlin
Tel.: +49 (0)30 42 85 10 90
Fax: +49 (0)30 42 85 10 92
INTERNET: http://www.logos-verlag.de

Foreword

Completing this dissertation was the biggest challenge in my life. Without the support of several people and parties, it would have been impossible to realize this manuscript. Hence at the very least through this foreword, they earn my deepest gratitude.

First of all, my thanks go to Deutscher Akademischer Austausch Dienst (DAAD) for the financial support during this project duration.

Afterwards, I would like to thank my first supervisor, Prof. Dr.-Ing. Hans-Michael Hanisch, for all inputs, discussion and direction, that were given to me for accomplishing this project. My gratitude is also addressed to Prof. Dr.-Ing. Dr. h. c. Joachim Ulrich, my second supervisor, who was interested in this project and who always gave the second recommendation for my scholarship extension.

Furthermore, I want to thank to the second reviewer namely Prof. Dr. Zbigniew Banaszak, who has given the second opinion to this thesis.

Likewise, I want to mention all my colleagues in our laboratory, Martin, Dirk, Sebastian, Hans-Christian, who provided valuable support by discussion, critics and correction of my work. My great thanks are also given to Christian, particularly for the discussion and the software tools, which were used in this work.

Both as well as my immediate family, namely my father, my mother, my father in law and my mother in law, thanks for all constant spirits and prayer. Most of all, my gratefulness is dedicated to my beloved wife Vivi as well as to my two beloved children, Akbar and Emier. Without their continuous support and prayer, I would not be sure if I was able to finish, what I have started.

M. Fauzan Amir
In July 2011

Acknowledgement

This thesis was printed by the support of the DAAD (Deutscher Akademischer Austausch Dienst/German Academic Exchange Service).

Content

List of Figures

List of Tables

1.   Introduction

Systematic and structured approaches have been adopted by the recipe-based standards i.e. ISA—S88.01 and the German Namur working group—NE33 [17,18], which are widely used by many chemical companies and the suppliers of process control systems for designing control strategies in batch processes. The recipe concept is actually a result of an intuitive design process based on the experience and non-formal knowledge of the process that must be controlled. However, it is not based on mathematical models. As a consequence, there is an unresolved issue of the formal method to analyze the control strategies to meet the desired specifications, before they are really implemented to real plants without resulting errors during the implementation.

On the other hand, the increasingly large literature in the batch scheduling area highlights the variety of successful applications of simulation, modeling and also analysis to discrete event systems [1-16]. However, there is still a significant gap between theory and practice. Mostly, the investigated cases use abstract models which usually neglect details of complexity of actual systems, with the aim of keeping the models simple. In fact such details are substantial for real implementation of the formal model of the controller. However, the formal methods must usually deal with the exponentially inherent complexity of the actual systems. Hence, efforts must be increasingly oriented toward the development of systematic, modular techniques for tackling the complexity of modeling and analysis of the real systems.

Hence, motivated by two things, i.e. the weakness of the experience approaches as well as the aim of closing the gap between theory and practice, particularly in designing a suitable control strategy for large systems using a well-known formal method, this thesis tries to develop some contributions inducted from the technical problems of an industrial-sized example. The Biodiesel batch-plants from a company in Indonesia are used as the application object to represent how the inherent complexity of the real system can be solved by the proposed formal methods. Summarily, the problems are explained as follows.

First of all, there is an increasing need for broadening the application scope of modeling methods with systematic and structured ways referring to the batch standards of Biodiesel batch-plants. In this case, the obtained results are particularly expected to be able to become a scientific reference for building the formal models of Biodiesel batch-plants efficiently, thus providing a significant impact to the design development of Biodiesel manufacturing in general.

Secondly, the method of analyzing a control strategy requires plant specifications which are defined formally. In this case, there are two desired plant behaviors. The first is the optimal behavior, which varies and depends on the considered economic factors. The appropriate criterion that directly corresponds to the production rate of a batch-plant are the waiting periods of process devices. The smallest waiting periods have a mean of high efficiency of the utilization of process devices in the plant strongly related to the accelerated production time. Whereas the desired second criterion is related to the process safety. As it is known, to minimize the waiting times in process units when they try to access conflict resources, a generally used approach is to refill the resources *as soon as possible after* the resource allocation. However, in real cases with long time constants, the cycle time of the process could be much longer than its allocation time, thus causing the idle times of the conflict resources in the metering tanks. Particularly for the dangerous resources, the idle times are exactly unwanted and must be able to be overcome by the suitable control strategy.

Thirdly, the modeling analysis of large plants usually deals with various complex problems, even for the case of a single process line. In this case, the design of the plant indicates that each process line can be driven by the basic recipe options. Although the result of an interview with the process engineer of the company stated that they ever tried to operate some combinations of the process recipe options, the unavailability of a clear (transparent) information of the sequence management of the process, however, raises doubts and some questions i.e. which option of the process can give the optimal result? How can the production time be improved? To answer such questions, the process options must be simulated and analyzed by means of a correct model. Nevertheless, the process options result in a huge model, even for each single production line, where the direct analysis to find an optimal path of the basic plant structure is time consuming and ineffective because of the complicated paths of the resulting model. Hence, there is a need to develop an effective way of getting rid of such complexity.

Fourthly, allocation of shared resources, for example: metering tanks, vacuum machines, stored tanks etc, is the crucial aspect of optimizing plant scheduling. The use of the exclusively shared resources likely arise conflicts. If the conflicts are not managed well by means of the scheduling system, the process will possibly deadlock, so it will possibly shut the plant down or even brings the plant into dangerous states. Otherwise the conflicts, solved by the correct control strategy, will maximize the plant throughput [5-9,16,19]. Nevertheless, all methods for determination of optimal resource allocation strategy—as a part of controller analysis for the whole system—suffer from the complexity problem, i.e. state explosion of the appropriate process models. Hence, there is a growing interest to develop an exact method to cope with the state explosion problem due to the coupling resources.

Fifthly, the problem of the state explosion is getting increased when a specific need of minimizing the idle time of the dangerous material in metering tanks must be solved while ensuring the optimal result. To cope with this problem, the refilling process of the dangerous substance into the metering tank must be performed *as close as possible before* the execution of the allocation, in order to minimize the idle times. By relying only on the experience approach, rather than obtaining the problem solving for the dangerous idle times, it is even difficult to provide the optimal solution for the complex system itself.

What are the contributions of this thesis?

The real system that is used in this work is a typical example for batch-plants. It therefore makes sense to explore approaches that can solve such problems and then are applied to this kind of systems. The general method of modeling itself is not new and has been applied for more than one decade in the field of recipe-based control. However, there are some major contributions which are developed from the following results, i.e.:

1. To broaden the application scope of the modeling of large systems especially for Biodiesel batch-plants by using the modular-based modeling method in accordance with the used batch standards.

2. To define formal specifications for waiting periods of the process devices as well as dangerous idle time, which both are used as objective functions for the desired controller analysis.

3. To propose an intuitive strategy to cope with the complexity of a huge model of each single production line due to its process options. The novel approach is based on the principle of the closest occupation time between consecutively used process devices.

4. To propose a systematic strategy based on a decomposition principle for tackling the state explosion problem caused by the complexity of resource allocation.

5. To propose an extra procedure how to minimize the idle time of dangerous resources with keeping ensuring the optimal allocation of the used resources.

6. To propose an alternative model expressing existing restrictions for the control strategy. The restrictions are required, for example, in the case of composing the two already optimized plant schedules into a single integrated system where the optimal behavior of each subsystem must remain unchanging when unified.

Outline of the Thesis

The contributions of the presented work are basically developed based on the general framework for model-based controller design which is commonly used in control engineering [37, 42, 43]. They are realized into chapters of this thesis constructing a formal method of modeling analysis as sketched in Figure 1.

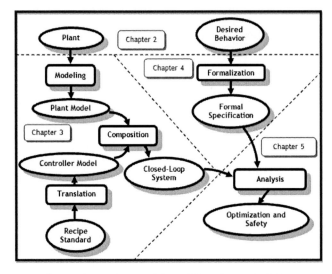

Figure 1. General framework for model-based controller design.

Chapter 2 discusses the plant description as well as the desired behaviors of the plants, mainly related to two production requirements, i.e. optimality and process safety. At first, the problem statement is addressed by the process design background why the recipe options are required and how the increased production speed can be achieved by managing the options of the process operation. The second problem that is described, are the conflicts in the resource allocation, not only the resources used exclusively by the process lines inside of each plant, but also the exclusively shared resource coupling all plants. The third problem is focused on the safety specification, i.e. a feature of the dangerous idle times which must be suppressed as little as possible for the purpose of a safe operation. These requirements are vitally important because they are needed to analyze a closed-loop system model which is built in the following chapter.

Chapter 3 presents a brief overview of Timed Net Condition/Event Systems (abbr. TNCES) as the used modeling method. Afterwards the plant functionalities are modularly modeled to form an open-loop model. The open-loop model is then joined with the recipe-based controller model into a closed-loop system model.

In Chapter 4, the desired behaviors of the plants are defined formally in the context of cyclic models of batch processes. The objective functions are waiting periods of process devices and dangerous idle times where the formally defined specifications are taken into account from the startup phase until the cyclic, stationary behavior.

As it has been predicted previously, the composed model of the real system has a huge size such that the model performance is very difficult to be analyzed. Hence, Chapter 5 proposes some strategies for tackling the complexity of the huge model. Starting with the complexity due to the process options, an approach which decomposes the original model (composition of the recipe options) and then reselects the applied basic recipes which only meet the closest occupation time between the consecutively used main process devices is proven to be able to overcome the complexity problem. Furthermore, the problem of the state explosion when analyzing each plant is tackled by a modular strategy using a refined original model. The refined model only contains the desired behavior of optimal subsystems. It is shown by simulation that the smaller behavior is contained by the refined model, such that the analysis can be done easier. Additionally, this chapter also demonstrates the strategy of minimizing the dangerous idle times in the state explosion. By shifting the dangerous resource refilling coinciding before the resource allocation and including it then in the refined model, zero idle time is reached successfully. Finally, a way to unify the optimal control strategy of each plant into a single integrated scheduling system is proposed. The strategy contains restrictions that ensure, that the optimal behavior of the composed systems is not changing at all (always fixed) after the unification. Evaluation of the performance of the whole plants shows that the proposed strategies are successfully implemented to achieve the expected results, i.e. the drastically reduced dynamic graphs, cost benefits of the accelerated production times and the improved process safety as well.

Finally, Chapter 6 draws conclusions of the presented work.

2.   Plant Description and Desired Behavior

To analyze a correct controller of recipe-controlled batch-plants, related process descriptions and plant specifications must be well known. Knowledge of the process is obtained by means of interpretation of process manufacturing described by basic recipes, whereas the plant specifications are known by means of the desired plant behaviors which are in this case oriented toward the achievement of a safe operation and optimal production scheduling. Related to the achievement of the manufacturing requirements of the investigated process, there are three underlying problems found i.e.

- Determination of the suitable control recipes from the existing basic recipe options which meets a criterion of the minimal waiting periods of main process devices in each single production line.
- Optimal allocation either for the limited resources coupling more than one production line inside of each plant, or for a conflict resource coupling the both plants overall.
- The dangerous idle times in the metering tanks that must be minimized while ensuring the optimal behavior of the plants.

These problems will be pointed out in the next sections started with the description of the used batch-plants.

2.1   General Plant Description

As shown in Figure 2, the two batch-plants that are used in this work are obviously flexible plants where the original design of the plants was not dedicated for the Biodiesel production. When market demand of Biodiesel was high, the original plants were then modified to have a capability for producing Biodiesel. Therefore, the equipment units installed in these plants are not similar and some parts are not meeting the actual design of a biodiesel batch-plant. In addition, the plants are operated depending on the market demand and the raw material supply of Biodiesel, or in other words the Biodiesel factory will be only operated when needed (by request).

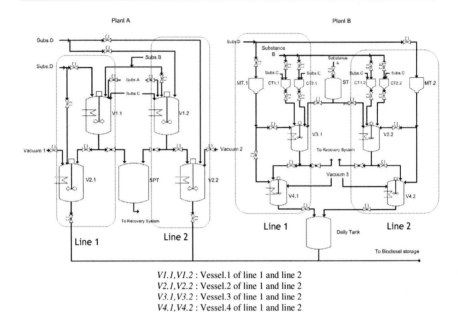

V1.1,V1.2 : Vessel.1 of line 1 and line 2
V2.1,V2.2 : Vessel.2 of line 1 and line 2
V3.1,V3.2 : Vessel.3 of line 1 and line 2
V4.1,V4.2 : Vessel.4 of line 1 and line 2

Figure 2. Process Flow Diagram (PFD) of the Biodiesel batch-plants.

2.1.1 Description of Plant A

Plant A sketched at the left side of Figure 2 consists of two similar production lines. Each line mainly consists of two vessels which are denoted with $V1.n$ and $V2.n$ ($n \in \{1, 2\}$ is the number of the line). Raw materials A, B and C are fed into the vessel $V1.n$ through pipelines connected to the storage tanks of the materials. Each pipeline is equipped with a flow meter to ensure an exact amount of the raw material supplied. Substance D for washing processes is also supplied into the vessels $V1.n$ and $V2.n$ through the separated pipelines which are equipped with a flow meter. Vacuum machines are only installed on vessels $V2.n$. The vacuum machines are needed for drying processes to ensure that the Biodiesel specification is fulfilled. Both vessels ($V1.n$ and $V2.n$) are facilitated by a heater and an agitator to be able to perform stirring and heating. A buffer tank SPT is used for storing a secondary product that is produced during the Biodiesel manufacturing.

Additionally, the tank capacity of SPT is designed for 40 MT (Metric Tons), the secondary product is produced in about 6.4 MT/batch and the further treatment of the secondary product (in another process) has a flow capacity of 6.4 MT/80 time units. The production capacity of this plant is maximal 28.5 MT Biodiesel/Batch where each product is directly transferred to the final storage tank with a flow capacity of 28.5 MT/102 time units.

2.1.1 Description of Plant B

Plant B sketched at the right side of Figure 2 also consists of two identical production lines. Each line mainly consists of two vessels i.e. vessel $V3.n$ and vessel $V4.n$ where n $\in$ {1, 2} is the number of the process line. Before feeding all the materials (A, B and C) into the main vessels and to ensure the exact amount of the reaction needs, the materials must be dosed in the appropriate metering tanks i.e. substance A into ST, substance B and followed manually by substance C into the consecutive pair of $CT1.n$ and $CT2.n$ and substance D into $MT.n$ as well. Then, the main reaction is started by evacuating all the substances A, B and C from the appropriate metering tanks into vessels $V3.n$ through the connected pipelines. Substance D is later evacuated from $MT.n$ into vessels $V3.n$ or vessel $V4.n$ for the purpose of a washing process. A vacuum machine is connected to the vessels $V4.1$ and $V4.2$, which is required for the exclusively drying process. The main vessels ($V3.n$ and $V4.n$) are facilitated by a heater and an agitator to be able to perform the jobs as stirring and heating. The tote tanks (tanks with 1 MT capacity) are used for storing the secondary product of the Biodiesel manufacturing that will be processed further by other units.

Additionally, the production capacity of the plant is maximal 10.6 MT Biodiesel/Batch. In opposition to Plant A, that transfers its products to the final storage tank directly, this plant uses daily tank DT as a buffer tank (temporarily) for its products in case of avoiding conflicts with Plant A when accessing the same pipeline heading to the final storage tank. The capacity of the daily tank (the buffer tank) is designed for 86 MT, while the pump capacity for transferring the products from the daily tank to the final storage is maximal 10.6 MT/25 time units.

2.2 Basic Recipe Options for Both Plants

A preliminary investigation on each single process line of those plants has given possibilities to improve the production speed by means of managing locations of washing processes of the Biodiesel. As it is known, one of the steps of manufacturing Biodiesel is a washing process, which is intended to remove any remaining catalyst, soap, salt, methanol, or free glycerol from the Biodiesel [20]. Based on the data of the used basic recipes, Plant A operates two times of washing processes to purify the Biodiesel, while Plant B requires three sequentially washed Biodiesel. In fact, all the main vessels in the Biodiesel plants are equipped with pipelines for accessing substance D (the washing material), such that the washing processes can be performed in any vessel along the production line, as long as the resource is available. Placing of the washing location influences the duration of the occupation of the main process devices, which can affect the waiting periods of the used process devices. In other words, the minimum waiting periods (maximizing the plant throughput) are a result from the efficiently occupied devices by the optimal control strategy. Shortly, the options of the washing location in each plant are summarized in the following tables.

Table 1 proposes three location options for the washing processes in the first plant. The first option states that the first washing (WS1) and the second washing (WS2) are both performed sequentially only in vessel $V2.n$. Otherwise the second option implies all washing processes (WS1 and WS2) performed only in vessel $V1.n$. The third option proposes the distributed locations of the washing processes, i.e. the first washing (WS1) performed in vessel $V1.n$ and the remaining one (WS2) finished in vessel $V2.n$.

Table 1. The washing locations in Plant A.

Plant A	Vessel V1.n	Vessel V2.n
Option 1	No	WS1, WS2
Option 2	WS1, WS2	No
Option 3	WS1	WS2

WS: washing

Table 2 represents the location options of the washing processes in the second plant. The first option offers the two washings (WS1 and WS2) performed consecutively in vessel $V3.n$, while the last one (WS3) is finished in vessel $V4.n$. The second option states that the first washing (WS1) is started in vessel $V3.n$, while the remaining washings (WS2, WS3) are finished in vessel $V4.n$. In the third option, all washing processes are only performed in vessel $V3.n$, otherwise the fourth option implies that only vessel $V4.n$ is used as the exclusive location for all washing processes.

Table 2. The washing locations in Plant B.

Plant B	Vessel V3.n	Vessel V4.n
Option 1	WS1,WS2	WS3
Option 2	WS1	WS2, WS3
Option 3	WS1,WS2, WS3	No
Option 4	No	WS1, WS2, WS3

WS: washing

As a consequence of the washing location options, the used basic recipes are expanded onto some appropriate basic recipe options. The expansion is intended to provide comprehensive information of the plants concerning the optimality aspects that shall be explored further on. Summarily, the first option of the washing locations in Plant A is realized by the basic recipe option given in Table 3, while the second and the third one are represented by Table 4 and Table 5, respectively. On the other hand, the four options of the washing locations in Plant B are completely realized by the basic recipe options that are provided in Table 6, Table 7, Table 8 and also Table 9. Principally, there are no significant differences among the steps underlying the basic recipe options, except for the location where the washing processes are placed.

Table 3. The first basic recipe option for Plant A.

Step.n	Process steps	Durations (tu)
S1	Supplying the raw materials into V1.n	80
S2	Stirring and heating V1.n	74
S3	Still stirring and keeping the desired temperature	90
S4	Cooling down V1.n*	51
S5	Discharging SP from V1.n into the SPT	65
S6	Transferring SP from SPT to the unit recovery	80
S7**	V1.n is emptied by transferring the crude BD into V2.n	36
S8	Adding D along with the stirring for Washing 1 in V2.n	15
S9	The process settles for some time (Settling 1***)	30
S10	Discharging D from V2.n	47
S11	Adding D along with the stirring for Washing 2 in V2.n	11
S12	The process settles for some time (Settling 2***)	30
S13	Discharging D from V2.n	15
S14	Heating and stirring V2.n	109
S15	Still stirring V2.n (keeping the temperature) and drying	177
S16	Cooling down V2.n and stopping the stirring	49
S17	V2.n is emptied by transferring BD to the final storage tank	102

* : by stopping the heating and the stirring for some time and then resulting in a distinct layer between BD on the top and the secondary product on the bottom.

** : the process transfer before all washings are performed in V2.n.

*** : resulting in a distinct layer with BD on the top and substance D on the bottom.

tu : time units, BD: Biodiesel, SP: secondary product, SPT: secondary product tank

Table 4. The second basic recipe option for Plant A.

Step.n	Process steps	Durations (tu)
S1	Supplying the raw materials into V1.n	80
S2	Stirring and heating V1.n	74
S3	Still stirring and keeping the desired temperature	90
S4	Cooling down V1.n*	51
S5	Discharging SP from V1.n into the SPT	65
S6	Transferring SP from SPT to the unit recovery	80
S7	Adding D along with stirring for Washing 1 in V1.n	15
S8	The process settles for some time (Settling 1***)	30
S9	Discharging D from V1.n	47
S10	Adding D along with the stirring for Washing 2 in V1.n	11
S11	The process settles for some time (Settling 2***)	30
S12	Discharging D from V1.n	15
S13**	V1.n is emptied by transferring the crude BD into V2.n	36
S14	Heating and stirring V2.n	109
S15	Still stirring V2.n (keeping the temperature) and drying	177
S16	Cooling down V2.n and stopping the stirring	49
S17	V2.n is emptied by transferring BD into the final storage tank	102

* :	by stopping the heating and the stirring for some time and then resulting in a distinct layer between BD on the top and the secondary product on the bottom.
** :	the process transfer after all washings are performed in V1.n.
*** :	resulting in a distinct layer with BD on the top and substance D on the bottom.
tu :	time units, BD: Biodiesel, SP: secondary product, SPT: secondary product tank

Table 5. The third basic recipe option for Plant A.

Step.n	Process steps	Durations (tu)
S1	Supplying the raw materials into V1.n	80
S2	Stirring and heating V1.n	74
S3	Still stirring and keeping the desired temperature	90
S4	Cooling down V1.n*	51
S5	Discharging SP from V1.n into the SPT	65
S6	Transferring SP from SPT to the unit recovery	80
S7	Adding D along with the stirring for Washing 1 in V1.n	15
S8	The process settles for some time (Settling 1***)	30
S9	Discharging D from V1.n	47
S10**	V1.n is emptied by transferring the crude BD into V2.n	36
S11	Adding D along with the stirring for Washing 2 in V2.n	11
S12	The process settles for some time (Settling 2***)	30
S13	Discharging D from V2.n	15
S14	Heating and stirring V2.n	109
S15	Still stirring V2.n (keeping the temperature) and drying	177
S16	Cooling down V2.n and stopping the stirring	49
S17	V2.n is emptied by transferring BD into the final storage tank	102

*   :   by stopping the heating and the stirring for some time and then resulting in

         a distinct layer between BD on the top and the secondary product on the bottom.

** :   the process transfer after washing 1 in V1.n and before washing 2 in V2.n.

*** :   resulting in a distinct layer with BD on the top and substance D on the bottom.

tu :   time units, BD: Biodiesel, SP: secondary product, SPT: secondary product tank

Table 6. The first basic recipe option for Plant B.

Step.n	Process steps	Dur. (tu)
S1	Dosing the raw materials (the substances A,B,C,D) into the metering tanks:	
	> substance A into ST	120
	> substance B into CT1.n	8
	> substance B into CT2.n (after CT1.n)	7
	> adding substance C into CT1.n (after substance B) along with mixing	5
	> adding substance C into CT2.n (after substance B) along with mixing	5
	> substance D into MT1 or MT2	5
S2	Evacuation of all materials from the metering tanks (ST,CT's and MT's) into V3.n	20
S3	Mixing and heating V3.n until the desired temperature is reached	60
S4	Keeping the desired temperature and still stirring	120
S5	Cooling down V3.n entering a settling process (Settling P)*	75
S6	Discharging the secondary product from V3.n to the tote tanks**	45
S7	Adding substance D along with the stirring for Washing 1 in V3.n	7
S8	Stopping the stirring and the process settles for some time (Settling 1****)	45
S9	Discharging substance D from V3.n	45
S10	Adding substance D along with the stirring for Washing 2 in V3.n	7
S11	Stopping the stirring and the process settles for some time (Settling 2****)	30
S12	Discharging substance D from V3.n	20
S13***	V3.n is emptied by transferring the crude BD into V4.n	20
S14	Adding substance D along with the stirring for Washing 3 in V4.n	7
S15	Stopping the stirring and the process settles for some time (Settling 3****)	30
S16	Discharging substance D from V4.n	20
S17	Heating and stirring V4.n until reaching the desired temperature	90
S18	Keeping the temperature, drying and stirring V4.n	90
S19	Cooling down V4.n and stopping the stirring	120
S20	Discharging BD from V4.n to the daily tank (temporary storage)	30
S21	Transferring BD from the daily tank to the final storage	25

* :	by stopping the heating and the stirring for some time and then resulting in a distinct layer between BD on the top and the secondary product on the bottom.	
** :	tanks with 1 MT capacity.	
*** :	the process transfer after washing 2 in V3.n.	
**** :	resulting in a distinct layer with BD on the top and substance D on the bottom.	
tu :	time units, BD: Biodiesel.	

Table 7. The second basic recipe option for Plant B.

Step.n	Process steps	Dur. (tu)
S1	Dosing the raw materials (the substances A,B,C,D) into the metering tanks:	
	> substance A into ST	120
	> substance B into CT1.n	8
	> substance B into CT2.n (after CT1.n)	7
	> adding substance C into CT1.n (after substance B) along with mixing	5
	> adding substance C into CT2.n (after substance B) along with mixing	5
	> substance D into MT1 or MT2	5
S2	Evacuation of all materials from the metering tanks (ST,CT's and MT's) into V3.n	20
S3	Mixing and heating V3.n until the desired temperature is reached	60
S4	Keeping the desired temperature and still stirring	120
S5	Cooling down V3.n entering a settling process (Settling P)*	75
S6	Discharging the secondary product from V3.n to the tote tanks**	45
S7	Adding substance D along with the stirring for Washing 1 in V3.n	7
S8	Stopping the stirring and the process settles for some time (Settling 1****)	45
S9	Discharging substance D from V3.n	45
S10***	V3.n is emptied by transferring the crude BD into V4.n	20
S11	Adding substance D along with the stirring for Washing 2 in V4.n	7
S12	Stopping the stirring and the process settles for some time (Settling 2****)	30
S13	Discharging substance D from V4.n	20
S14	Adding substance D along with the stirring for Washing 3 in V4.n	7
S15	Stopping the stirring and the process settles for some time (Settling 3****)	30
S16	Discharging substance D from V4.n	20
S17	Heating and stirring V4.n until reaching the desired temperature	90
S18	Keeping the temperature, drying and stirring V4.n	90
S19	Cooling down V4.n and stopping the stirring	120
S20	Discharging BD from V4.n to the daily tank (temporary storage)	30
S21	Transferring BD from the daily tank to the final storage	25

*   :   by stopping the heating and the stirring for some time and then resulting in
          a distinct layer between BD on the top and the secondary product on the bottom.
**   :   tanks with 1 MT capacity.
***   :   the process transfer after washing 1 in V3.n.
**** :   resulting in a distinct layer with BD on the top and substance D on the bottom.
tu   :   time units, BD: Biodiesel.

Table 8. The third basic recipe option for Plant B.

Step.n	Process steps	Dur. (tu)
S1	Dosing the raw materials (the substances A,B,C,D) into the metering tanks:	
	> substance A into ST	120
	> substance B into CT1.n	8
	> substance B into CT2.n (after CT1.n)	7
	> adding substance C into CT1.n (after substance B) along with mixing	5
	> adding substance C into CT2.n (after substance B) along with mixing	5
	> substance D into MT1 or MT2	5
S2	Evacuation of all materials from the metering tanks (ST,CT's and MT's) into V3.n	20
S3	Mixing and heating V3.n until the desired temperature is reached	60
S4	Keeping the desired temperature and still stirring	120
S5	Cooling down V3.n entering a settling process (Settling P)*	75
S6	Discharging the secondary product from V3.n to the tote tanks**	45
S7	Adding substance D along with the stirring for Washing 1 in V3.n	7
S8	Stopping the stirring and the process settles for some time (Settling 1****)	45
S9	Discharging substance D from V3.n	45
S10	Adding substance D along with the stirring for Washing 2 in V3.n	7
S11	Stopping the stirring and the process settles for some time (Settling 2****)	30
S12	Discharging substance D from V3.n	20
S13	Adding substance D along with the stirring for Washing 3 in V3.n	7
S14	Stopping the stirring and the process settles for some time (Settling 3****)	30
S15	Discharging substance D from V3.n	20
S16***	V3.n is emptied by transferring the crude BD into V4.n	20
S17	Heating and stirring V4.n until reaching the desired temperature	90
S18	Keeping the temperature, drying and stirring V4.n	90
S19	Cooling down V4.n and stopping the stirring	120
S20	Discharging BD from V4.n to the daily tank (temporary storage)	30
S21	Transferring BD from the daily tank to the final storage	25
* :	by stopping the heating and the stirring for some time and then resulting in a distinct layer between BD on the top and the secondary product on the bottom.	
** :	tanks with 1 MT capacity.	
*** :	the process transfer after all washings are performed in V3.n.	
**** :	resulting in a distinct layer with BD on the top and substance D on the bottom.	
tu :	time units, BD: Biodiesel.	

Table 9. The fourth basic recipe option for Plant B.

Step.n	Process steps	Dur. (tu)
S1	Dosing the raw materials (the substances A,B,C,D) into the metering tanks:	
	> substance A into ST	120
	> substance B into CT1.n	8
	> substance B into CT2.n (after CT1.n)	7
	> adding substance C into CT1.n (after substance B) along with mixing	5
	> adding substance C into CT2.n (after substance B) along with mixing	5
	> substance D into MT1 or MT2	5
S2	Evacuation of all materials from the metering tanks (ST,CT's and MT's) into V3.n	20
S3	Mixing and heating V3.n until the desired temperature is reached	60
S4	Keeping the desired temperature and still stirring	120
S5	Cooling down V3.n entering a settling process (Settling P)*	75
S6	Discharging the secondary product from V3.n to the tote tanks**	45
S7***	V3.n is emptied by transferring the crude BD into V4.n	20
S8	Adding substance D along with the stirring for Washing 1 in V4.n	7
S9	Stopping the stirring and the process settles for some time (Settling 1****)	45
S10	Discharging substance D from V4.n	45
S11	Adding substance D along with the stirring forWashing 2 in V4.n	7
S12	Stopping the stirring and the process settles for some time (Settling 2****)	30
S13	Discharging substance D from V4.n	20
S14	Adding substance D along with the stirring for Washing 3 in V4.n	7
S15	Stopping the stirring and the process settles for some time (Settling 3****)	30
S16	Discharging substance D from V4.n	20
S17	Heating and stirring V4.n until reaching the desired temperature	90
S18	Keeping the temperature, drying and stirring V4.n	90
S19	Cooling down V4.n and stopping the stirring	120
S20	Discharging BD from V4.n to the daily tank (temporary storage)	30
S21	Transferring BD from the daily tank to the final storage	25

* :	by stopping the heating and the stirring for some time and then resulting in a distinct layer between BD on the top and the secondary product on the bottom.	
** :	tanks with 1 MT capacity.	
*** :	the process transfer before all washings are performed in V4.n.	
**** :	resulting in a distinct layer with BD on the top and substance D on the bottom.	
tu :	time units, BD: Biodiesel.	

Since each plant consists of two identical production lines, the basic recipe options above must be then duplicated to represent the actual size of all plants. Therefore the totally composed plants contain fourteen recipe options where six recipe options are dedicated for the first plant while eight recipe options are offered for the second plant. To provide the global model, the recipe options have to be interconnected by the used resources as mapped in Figure 3. *TRA* denotes the limited resources shared by the production lines inside of the first plant, while *TRB* represents the exclusively shared resources by the production lines within the second plant. Both plants are finally unified by one together used resource i.e. *TRC*. Clearly, the following will discuss the description of the used resources of all plants, particularly the restrictions existing in the resource allocation.

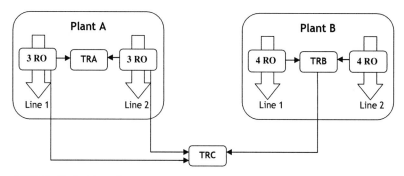

3 RO : three basic recipe options
4 RO : four basic recipe options
TRA : resources shared by two production lines in plant A
TRB : resources shared by two production lines in plant B
TRC : a resource coupling the both plants

Figure 3. Mapping on the use of the limited resources in the batch-plants.

2.3   Conflict Resources

In this case, constraints of the use of the limited resources existing in both plants are described as follows.

2.3.1   Resource Constraints in Plant A (TRA)

[1]   The availability of the pipeline allocation for accessing substance *B* is an exclusively shared resource. Substance *B* is not allowed to be simultaneously accessed by two production lines of the plant because there is only one flow meter available on the

associated pipelines. The flow meter is required to ensure the exact amount of feeding Substance B into the main vessels.

[2]   Since there is only one flow meter installed on the pipeline for accessing substance D, the availability of the pipeline allocation is an exclusively shared resource too.

[3]   The products of each process line must be directly transferred to the final storage in another process unit (not part of the plant) after ensuring the free status of the pipeline which will be used.

2.3.2 Resource Constraints in Plant B (TRB)

[1]   Substance A is exactly dosed into the metering tank ST and then each process line allocates the metering tank ST exclusively.

[2]   Filling (refilling) substance B followed thereafter by substance C into the consecutive pair of the metering tanks $CT1.n$ and $CT2.n$ must be performed exclusively for each tank.

[3]   Filling (refilling) substance D into the metering tanks $MT1$ and $MT2$ is not allowed to be performed simultaneously.

[4]   The dose of substance D in the metering tanks $MT1$ or $MT2$ is only allocated exclusively for one of the vessels, $V3.n$ or $V4.n$.

[5]   The vacuum machine is used exclusively for vessel $V4.1$ or vessel $V4.2$.

[6]   To ensure the exact amount of each resulting product, an evacuation of the product from vessels $V4.1$ or vessel $V4.2$ into the daily tank DT must be done exclusively. Additionally, the daily tank must have the adequate capacity before the transfer is realized.

2.3.3 A Conflict Resource Coupling Both Plants (TRC)

As it is shown at the downstream area of the process flow diagram in Figure 2, there is a resource coupling two plants, namely the availability of the pipeline used for transferring the Biodiesel product from two plants to a final storage tank. As mentioned before, the products of Plant A are transferred directly to the final storage tank, otherwise the Plant B uses the buffer tank (the daily tank DT) temporarily and then storing the product permanently after checking whether the final line is already "free-job" and ready for use.

2.4   Dangerous Resource

Particularly for substance B that is categorized as a dangerous material, the idle times of this material particularly on the metering tanks $MT.1$ and $MT.2$ in the second plant (Plant B) must be suppressed as small as possible, but keeping ensuring the optimal result. In other words, the desired control strategy is not only containing the conflict resolution for the optimal resource allocation, but also has to be able to minimize the dangerous idle times existing in the plant.

3.   Modeling of the Batch-Plants

If one considers the work of Hanisch et al in [8, 9], timed-arc Petri nets were used as formal models particularly for modeling the plant description of batch processes. For more information about Petri nets, the interested readers are referred to references [21-25]. Although Petri nets is a powerful mathematical and graphical tool for modeling hierarchical designs, there are lacks of modularity and composition properties which precisely constitute basic principles of designing a process control for large plants that requires systematical and structured ways, as adopted by the widely used batch standards. Hence, originated by the work of Sreenivas and Krogh that proposed the concept of condition and event signals [26], a modular-based method i.e. Condition/Event Systems (NCES) and its extension to describe time behavior i.e. Timed Net Condition/Event systems (TNCES), has been developed [7,27,40].

Successful applications of the (T)NCES method to batch processes and in other contexts of manufacturing industry can be easily found in a number of publications, e.g. in [6, 7, 28-37, 40, 41]. Although basically TNCES are not Petri nets, TNCES take advantage of the usage of graphical representations of Petri nets, such that it will be more understandable by engineers intuitively. Additionally, the timing concept of TNCES provides a realistic model which is very close to the actual plant behavior, such that the method is not only promising an approach that is well proven in the theoretical point of view but also strongly sufficient to solve the actual problems arising from real large systems with high complexity. Nevertheless, more work is still needed before the method can be really put into practice.

As is seen in Figure 4, common elements in TNCES use the same symbols as timed-arc Petri nets unless signal arcs are added to represent the interconnection among modules and the nets within the modules. Condition arcs are graphically represented by arcs with a black dot instead of the arrowhead that emits state information of places in a module that is used to enable the connected transitions in other modules. Otherwise, inhibitor arcs are symbolized by a blank dot, instead of the arrowhead, that has an opposite function with respect to condition arcs, where the contained signals transmit the information for disabling the connected transitions in other modules. In addition, event signals denoted by arcs with zigzag symbols provide information about state transitions of a module.

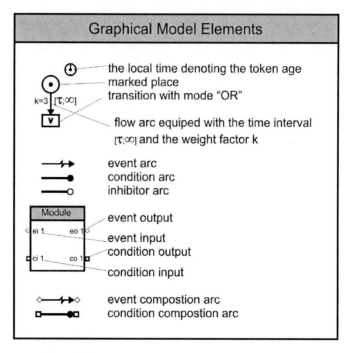

Figure 4. Graphical model elements of TNCES (see reference [7]).

To summarize the firing rules of TNCES, one can see three rows in Figure 5. The first row shows that the enabled transition (t1) is forced to fire by an event input (ei) which has a value "true" (denoted by bold symbols). The event input is related to a firing transition in another module. The second row shows that firing transition (t1) is enabled by the true value of the condition input (ci). The condition input is related to the status of a marked place in another module. The third row shows the transition (t1) with a pre-arc annotated with a time interval, which fires according to the earliest firing rule [8].

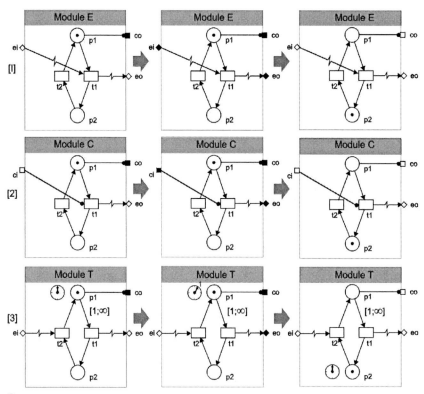

⊕ : Clock symbols (denoting the tokens age)

Figure 5. Firing rules of TNCES (see reference [7]).

To analyze the model dynamics, there is a need to compose modules into a new-flat module intuitively by "gluing" condition connections and event connections to condition arcs and event arcs of the composed modules. Completely, the definitions of the composition have been described in [27]. The model dynamics can be analyzed by means of a state graph called dynamic graph. The dynamic graph contains sequences of all possible firing transitions and all reachable states representing marking positions and also the time of the marked places of the composed modules.

Software tools called TNCES-Editor and TNCES-Workbench, that serve as a model editor, dynamic graph calculation as well as an evaluating tool have been developed by the host group i.e. the Chair of Automation Technology, Institute for Computer Science, Martin Luther University of Halle-Wittenberg, Germany [38, 41]. The software tools input models graphically and automatically visualize appropriate dynamic graphs with the additional features such as *find states, dead states, Gantt chart* etc., which are important for evaluation and performance analysis of the modeled systems.

The contribution of this section focuses on the broadening of the application scope of the TNCES-based modeling, which is addressed to the Biodiesel batch-plants following the systematic and structured ways as recommended by the batch standards. Principally, the recipe-based concept refers to simplification of complex plants by constructing the control structure of the plant into several stages and modules. By following the systematic guidelines which were developed in [7], the control structure composing models of the plants and models of the recipe controller can be designed hierarchically and modularly for the needs of controller analysis of the plants.

3.1   Control Structure

The control structure which becomes the reference for building the plant models modularly is displayed in Figure 6.

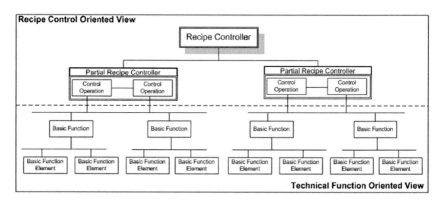

Figure 6. Control structure referring to the batch standards (see reference [7]).

In general, there are two orientations classifying the control structure as shown by a dashed line dividing Figure 6 into two parts. Both parts of the block diagram represent a closed-loop model between the uncontrolled plant and its controller. The bottom part describes the technical function oriented view which places the plant as an open-loop system (uncontrolled). Here, the plant process functionalities are defined and structured. Atomic elements representing the lowest level at this hierarchy are called Basic Function Element (BFE) which are responsible for defining a single actuation for plant actuators such as pumps, valves etc. In other words, this level realizes the control orders to all technical equipments of the batch-plants. However, in the context of the functionality described in this work, modeling of the atomic action control is not necessary, since the basic process functionalities are adequately represented by the upper control level, i.e. basic functions. Indeed, the detailed model of the basic function elements exactly increases computational effort in analysis. Here, the inclusion of the basic function elements in the control structure is only done for completeness. Hence, this level is no longer used in the sequel.

Elementary process tasks such as heating, dosing, stirring etc are represented by basic functions (BF) at the next level. The operation instructions are the discrete events which may be time events caused by clocks or state events caused by the reaching of a threshold of continuous variables such as level, temperature and so on. Basic functions can independently run parallel and do not depend on any hierarchical orders. Basic functions are not product-specific and thus are usable for different recipes.

The top part of Figure 6 describes the recipe control oriented view which structures a recipe control into three hierarchical levels. The set of basic functions defining the algorithm of a particular process unit is called basic function sequence (BFS) and is controlled by a control operation (CO) at the basic level in the recipe hierarchy. At the upper control level, control operations run consecutively forming an algorithm of the subprocess called by partial recipe control (PR). The partial control recipes composed into a sequence forms then a control recipe module (CR) which constitutes the highest level in this control structure to represent the whole process recipe for producing a specific product in a batch-plant.

Unfortunately, resource allocation which plays an exactly vital role in optimizing a scheduling system, is not explicitly described, neither in recipe controller nor in the plant description in the batch standards. Practically, the resource allocation strategy is usually designed being based on intuitive approaches of operational staff or controller programmers which of course cannot guarantee the exact resolutions over the happening conflicts and moreover to achieve an optimal criterion. Even in worse scenarios, human errors in the handling of the resource allocation problem cause a shutdown of the plants or other dangerous situations. Hence, in order to synthesize the optimal control strategy, the realized models have to include the resource units (TR) in their control structure. The resource allocation strategy must be determined by the recipe control level which serves as relatively autonomous process units of the automation structure, namely the control operation (CO).

The block diagram in Figure 7 represents the realization of the modeling concept of the recipe-based control and the plant description including the resource element in the automation structure. Signals of events and conditions realize a hierarchically feedback control scheme among the linked modules in connection with achieving the desired set points. Simply, an event signal denotes a discrete set point assigned to a lower level module related to a command of execution. The executions may be: starting, stopping, allocating as well as deallocating of the modules. While the condition signal serves as the status information provider of the controlled module, it is indicating whether it is available to be executed by the upper module level. Clearly, interpretation of each signal is given by Table 10.

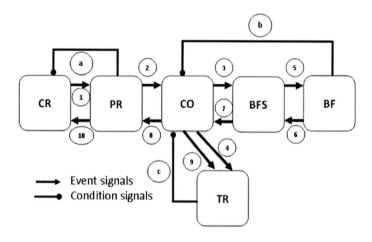

Figure 7. Realized control structure (see reference [7]).

Table 10. Interpretation of signals in the control structure.

Signals	Remarks
Event 1	To start the production process, the control recipe module (CR) sends the starting signals to one or more partial recipe modules (PR) as defined in the production schedule.
Event 2	The partial control recipe module (PR) activates the first control operation (CO) in its control operation sequence.
Event 3&4	After ensuring the availability of the appropriate technical resources (TR, see cond. signal c), the control operation (CO) allocates the needed technical resources (TR) and then starts its basic function sequence (BFS).
Event 5	The first basic function (BF) in the basic function sequence module (BFS) is started. The basic function operates the technical function of the plant.
Event 6	The basic function (BF) ends its task and sends the stop signal back to its basic function sequence (BFS).
Event 7	After all jobs of the basic functions (BF) are finished, the basic function sequence (BFS) terminates its operation and sends the stop signal back to its control operation (CO).
Event 8&9	As soon after receiving the stop signal from the basic function sequence (BFS), the control operation (CO) will enter the end state and send the stop signal to its partial recipe (PR). In this state, the control operation (CO) deallocates the used technical resources (TR).
Event 10	After finishing the sequence of the control operations, the partial recipe (PR) sends the stop signal to the control recipe (CR). The control recipe (CR) ends when all associated partial recipes (PR) have finished their tasks.
Condition a	This condition signal represents the availability of the partial recipe (PR) to be started by the corresponding control recipe (CR).
Condition b	Before starting the basic function sequence (BFS), by means of these signals the control operation (CO) must check the free status of the underlying basic functions (BF), to ensure that the basic functions are available to be executed.
Condition c	Before starting the basic function sequence (BFS), the control operation (CO) must check the availability of the needed resources (TR) by means of these condition signals.

3.2 Implemented Control Structure

In the following, implementation aspects of the modeling of the batch-plants using a top-down approach are described. The top-down technique is stepwise refinement of a whole control scenario into its substructures by following the previously described control structure design. Each successive step contains an increasing detail of subnets or pre-designed modules until the most basic level of the control structure is reached. The one-by-one implemented modules would be presented on generic forms as described in the following.

3.2.1   Control Recipe

As the controller with the highest level, a control recipe module defines all production activities in a batch-plant. The control recipe module organizes the sequence of the sub-processes i.e. partial control recipes underlying a control algorithm for producing a product. As modeled in Figure 8, to start the first execution, the condition signal *PR1_INAC* reporting the available state of the first partial control recipe would be evaluated. If the module is available, the execution is realized by the event signal *START_PR1*. After receiving the feedback signal indicating the end of the execution, the state in the control recipe is then changed by moving the current token onto place *W_PR2,* namely the waiting state for subsequent execution. The same pattern is applied for all the controlled sequential steps until the current token reaches the last place *FINISH* and a cycle is eventually created.

With regard to implementation to the batch-plants, the process interpretation for defining the applied control recipe is described by taking into account the following considerations. As it is introduced in Chapter 2, one-step of the Biodiesel manufacturing is the washing process which is intended to purify Biodiesel. The fact shows that the washing locations can be placed in any vessel along the production line in the plants (see Table 1 and Table 2 in Chapter 2). Due to the fact of the washing location options, the basic recipes for manufacturing Biodiesel can be expanded onto several control recipe options as given in Table 11.

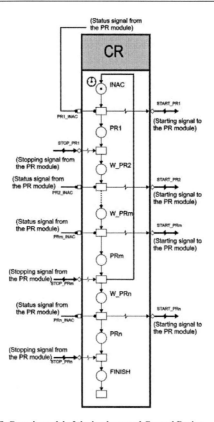

Figure 8. Generic model of the implemented Control Recipe.

Table 11. The control recipe options.

CR	Plant A (two sequential washings)	Plant B (three sequential washings)
CR1.n	The first option of the control recipe: WS1 and WS2 are all performed in V2.n. The controlled modules : PR1.n, PR2.n, PR3.n	The first option of the control recipe: WS1 and WS2 are both performed in V3.n whereas WS3 is performed in V4.n. The controlled modules : PR1.n, PR2.n, PR3.n.
CR2.n	The second option of the control recipe: WS1 and WS2 are otherwise performed in V1.n. The controlled modules : PR1.n, PR2.n, PR3.n.	The second option of the control recipe: while WS1 is performed in V3.n, WS2 and WS3 are otherwise performed in V4.n. The controlled modules : PR1, PR2.n, PR3.
CR3.n	The third option of the control recipe: WS1 is performed in V1.n whereas WS2 is performed in V2.n. The controlled modules : PR1.n, PR2.n, PR3.n.	The third option of the control recipe: WS1, WS2, WS are all performed in V3.n. The controlled modules : PR1, PR2.n, PR3.n.
CR4.n	No	The fourth option of the control recipe: WS1, WS2, WS are all performed in V4.n. The controlled modules : PR1, PR2.n, PR3.n.

WS: washing process, n: the line number.

3.2.2 Partial Control Recipe

To distribute the responsibility on a process control, the whole control scenario must be partitioned into relatively loosely connected modules, i.e. partial control recipes, which can be processed in relatively independent operation units. A partial control recipe module handles one or several unit operations encapsulated into the procedural control of the equipment units. Concerning on the application, the overall functionality provided by the batch-plants is represented by three partial recipes, i.e. the main reaction, the process purification and the finishing process. Table 12 defines the subprocesses. Transformation on the sequential orders to the first generic model of the partial control recipe can be seen in Figure 9.

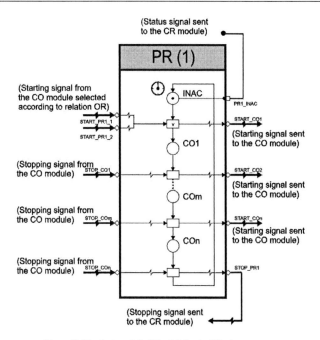

Figure 9. The first model of Partial Control Recipe.

It is typical for sequential function charts where the first step always begins with a state *START* by releasing the starting signal to the first control operation module. Each execution in the sequential function chart is terminated by the stopping signal received from each already accomplished control operation. Finally, after the completion of all the operation tasks, the partial control recipe sends the terminating signal to its control recipe module (upper level), to step up to a next state.

One of the advantages of the TNCES method is a feature of the transition mode OR, which enables the functionality of a module to be exclusively shared by other user modules. For example, two or more control recipe modules share a similar partial control recipe exclusively, as illustrated in Figure 9. The exclusive access is reflected by two (or more) input signals (sourced from two or more different control recipe modules) connected to the same transition via relation *OR*. It means, that only one signal would be followed up by the module at the same time. PR1.n and PR3.n are belonging to this kind of model. This concept also holds for other module types, as one can see in the upcoming subsections.

Table 12. The implemented partial control recipes.

PR	Plant A	Plant B
PR1.n	Performing main process reaction in V1.n, the controlled modules: CO1.n, CO2.n and CO3.n.	Performing main process reaction in V3.n, the controlled modules: CO1.n, CO2.n and CO3.n.
PR2.n	Performing process purification, the controlled modules: CO4.n,CO5.n, CO6.n and CO7.n	Performing process purification, the controlled modules: CO4.n, CO5.n. CO6.n, CO7.n and CO8.n.
PR3.n	Finishing process, the controlled modules: CO8.n.	Finishing process, the controlled modules: CO9.n.

n: the line number

Particularly for the process purification—denoted by PR2.n— which can be performed in the different locations, the corresponding partial control recipe must accommodate the options. They are configured by the independent, parallel branches that construct the sequential function chart, as exemplified by the second partial recipe model in Figure 10.

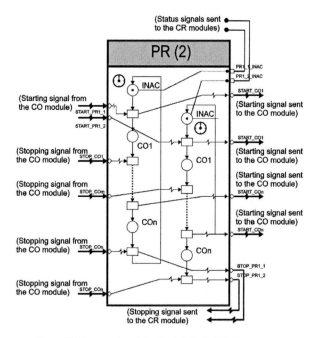

Figure 10. The second model of Partial Control Recipe.

3.2.3 Control Operation

By taking into consideration the availability of the needed resources and also the underlying basic functions, a control operation is responsible for determining when execution of a process unit in the plant must be started. As seen in the module Figure 11, when the executing order coming from the upper level is received, the token in place *INAC* must be removed onto the waiting place *INIT* for checking whether the underlying basic functions are ready to be assigned and the needed resources are already available. Once all the statuses are true, the operation may be started by residing the token into state *RUN* and by releasing a starting signal to the corresponding basic function sequence module. Then, all basic functions are executed following the algorithm of the basic function sequence. Once all tasks are completed, the used resources are soon deallocated and the token enter then place *END to* represent the final state. At this end state, the stopping signal is sent back to its partial recipe module (the upper level), to change the current operation state to the forthcoming operation order.

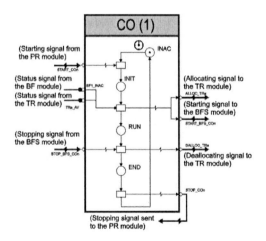

Figure 11. The first model of Control Operation.

To accommodate similar operation tasks, yet performed in different operation locations, a control operation module contains independent parallel nets which represent independently concurrent operations. The representative example is the washing process, namely CO4.n, CO5.n for Plant A and CO4.n, CO5.n, CO7.n for Plant B, which can be employed either in the first vessel or in the second vessel. Figure 12 models the appropriate behavior. Afterwards, the transformation on the real process units to the control operation modules is done according to the interpretations given in Table 13.

Table 13. The implemented control operations.

CO	Plant A	Plant B
CO1.n	Feeding all substances into V1.n. The controlled modules: BFS_CO1.n, BF1.n, TR1 (A), TR3 and TR4.	Transferring all substances from the metering tanks into V3.n. The controlled modules: BFS_CO1.n, BF1.n, TR1 (A), TR3 and TR4.
CO2.n	Heating, stirring and cooling V1.n. The controlled modules: BFS_CO2.n, BF2.n, BF3.n, BF4.n and BF5.n.	Heating, stirring and cooling V3.n. The controlled modules: BFS_CO2.n, BF2.n, BF3.n, BF4.n and BF5.n.
CO3.n	Discharging SP. The controlled modules : BFS_CO3.n, BF6.n and TR8.	Discharging SP. The controlled modules : BFS_CO3.n and BF6.n.
CO4.n	Washing 1. The controlled modules : BFS_CO4.n, BF7.n, BF8.n, BF9.n, BF10.n and TR5.	Washing 1. The controlled modules : BFS_CO4.n, BF7.n, BF8.n, BF9.n, BF10.n and TR5.
CO5.n	Washing 2. The controlled modules : BFS_CO5.n, BF11.n, BF12.n, BF13.n, BF14.n and TR5.	Washing 2. The controlled modules : BFS_CO5.n, BF11.n, BF12.n, BF13.n, BF14.n and TR5.
CO6.n	Transferring BD from V1n to V2.n. The controlled modules: BFS_CO6.n, BF15.n, TR1(D) and TR2 (A).	Transferring BD from V3.n to V4.n. The controlled modules: BFS_CO6.n, BF15.n, TR1(D) and TR2 (A).
CO7.n	Heating, drying and cooling V2.n. The controlled modules: BFS_CO7.n, BF16.n, BF17.n, BF18.n, BF19.n and TR6.	Washing 3. The controlled modules : BFS_CO7.n, BF16.n, BF17.n, BF18.n, BF19.n and TR5.
CO8.n	Storing Biodiesel directly into the final storage tank. The controlled modules: BF_CO8.n, BF20.n, TR2 (D) and TR7.	Heating, drying and cooling V4.n. The controlled modules: BFS_CO8.n, BF20.n, BF21.n, BF22.n, BF23.n and TR6.
CO9.n	-	Storing Biodiesel into the daily tank. The controlled modules: BF_CO9.n, BF24.n, TR2 (D) and TR7.

(A) : only allocation, (D): only deallocation, n : the line number (each plant consists of two lines), SP: the secondary product, BD: Biodiesel.

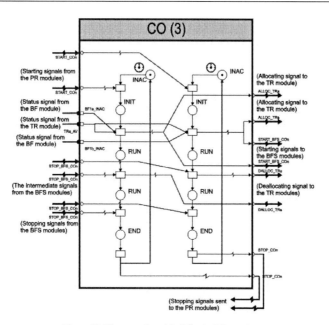

Figure 12. The second model of Control Operation.

3.2.4 Technical Resources

Modeling of technical resources is the most significant part of the plant modeling, since it is related to the primary aim of this work, i.e. the optimal allocation of the used resources. The modeling of the resource units follows the projection of the resource unit's functionality distinguished by [7, 17] into four resource types i.e.

Type 1: Buffer with internal refilling

This resource is typically a predecessor of the production equipment which is usually refilled periodically under an assumption of material sources being always available. Short periods of refilling are followed by long periods of allocation, or vice versa. The allocation type can be exclusive or not, depending on the technical restriction of the resources. As for example, metering tanks for dosing raw materials can be categorized into this type.

Type 2: Non refillable plant devices

Technical resources belonging to this kind of device are plant internal buffer resources, which are fed by a particular control operation and reloaded by a succeeding control operation. In many cases, the resource capacity and the resource connections (inlet or outlet) are the technical limitations of this resource type. All vessels or tanks in the plant are examples of this category.

Type 3: Ring mains

The resource is defined under an assumption of unlimited material source and therefore it does not need the refilling units. Examples of the resource type are process water, pumps, vacuum machines, heating steam systems etc. The allocation type can be exclusive or not, depending on the capacity restriction of the resources itself, for instance the actual performance of a pump limited by maximum flow etc.

Type 4: Buffer with external refilling

This buffer type is characterized by external refilling which is designed to be uncontrolled by an intra-process operation. As for example, refilling of solid catalysts must be performed manually by the plant staff.

In fact the four resource types above are encountered in the investigated batch-plants with some variant models. Summarily, interpretation on the used resources is given in Table 14.

Table 14. The implemented technical resources.

TR	Description (Plant A)	Description (Plant B)
TR1	Vessel V1.n: type 2, the capacity is not defined (always available).	Vessel V3.n: type 2, the capacity is not defined (always available).
TR2	Vessel V2.n: type 2, the capacity is not defined (always available).	Vessel V4.n: type 2, the capacity is not defined (always available).
TR3	Substance A: type 3, non-exclusively used connection points.	Metering tank ST for subs. A: type 1, exclusively used.
TR4	Substances B and C: type 3 and 4, exclusively used.	Metering tanks CT1.n & CT2.n for subs. B and C: type 1 and 4, exclusively used and exclusively refilled.
TR5	Substance D: type 3, exclusively used.	Metering tank MT.n for subs. D: type 1, exclusively used, exclusively refilled.
TR6	Vacuum machine: type 3, non-exclusively used connection points.	Vacuum machine: type 3, exclusively used connection points.
TR7	BD storage tank: type 2, undefined capacity, exclusively used connection points.	Daily Tank: type 2, with defined capacity, exclusively used connection points.
TR8	SP Tank: type 2, with defined capacity, non-exclusively used connection points.	BD storage tank: type 2, undefined capacity, exclusively used connection points.

BD: Biodiesel, SP: Secondary product

Resource Types 1, 3 and 4

The behavior complexity of resources Type 1 found in the plants ranges from the common type, i.e. the exclusive allocation, to the particular type requiring to be exclusively refilled by their source. The first model representing the exclusively allocated resource can be seen in Figure 13 with the initial state denoted by the marking place *REFILL*. After the refilling of the resource for τ time units, the resource is then available for allocation as indicated by the marking place *AVAILABLE* and all at once forwarding the true status to the condition output *TR_av*. By the transition mode *OR*, only one executing signal is accepted at the same time to allocate the resource. During the exclusive allocation, the token resides in place *USED* for some time, prior to finally return to the place *REFILL* for the next refilling. The application examples of resources type 1 are the metering tanks TR3, TR4 and TR5 that are used for dosing raw materials in the second plant, as already informed in Table 14.

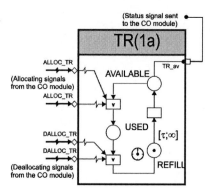

Figure 13. Model of resource Type 1 with exclusive allocation.

Particularly for the metering tanks MT1 and MT2 (TR5), beside of the restriction of the use of the resources (allocation), the other restriction is also found at the same inlet point of both metering tanks where the refilling of the two resources is not allowed to be done simultaneously. The model in Figure 14 meets the given specifications. As can be seen, the marking place *CON_AV* represents the availability of the inlet point for the refilling. While the waiting places *W1* and *W2* serve as buffer places added to prevent deadlocks when the refilling of the metering tanks is in conflict.

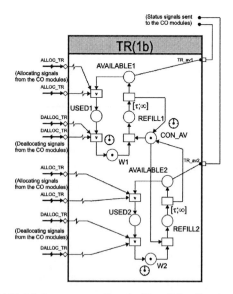

Figure 14. Model of two resources, Type 1, with exclusive allocation.

Another variance of resource behavior is also found in Plant B namely the combination of Type 1 and Type 4 all at once. Refilling of substance B followed by adding substance C manually into the consecutive pair of the metering tanks *CT1.n* and *CT2.n* is belonging to this kind of model. As shown by the corresponding model in Figure 15, the places *REFILL1(2)a* and *REFILL1(2)b* represent the exclusively refilling activity of resource type 1, i.e. metering tanks *CT1.n* and *CT2.n* with the inlet connection managed by the place *CON_AV,* whereas adding substance C manually is denoted by the subsequent places *REFILL1(2)c (MAN)* and *REFILL1(2)d (MAN)* representing the combination of resource Type 4.

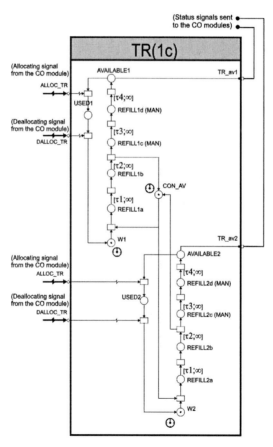

Figure 15. Model of combination of Type 1 and Type 4, with exclusive allocation as well as exclusive refilling.

In contrast to resource Type 1 which needs the refilling process, resources Type 3 (ring mains) do not need the refilling process. With the assumption of unlimited source capacity, this resource type neglects the need of the refilling. Supposing that raw materials are stored in very large storage tanks (see TR3, TR4, TR5 in Table 14, column Plant A), such resources can be included into this category. Likewise with the vacuum machines (TR6), which are assumed to be always available, they are also classified into this resource type. Figures 16 and Figure 17 model this kind of resource with two allocation types i.e. exclusively used connection points and non-exclusively used connection points respectively. As one can see, due to the assumption of the unlimited capacity, the refilling element is omitted in this model. Meanwhile, to represent the independency in the use of resources (non-exclusive), more than one token reside in the place *AVAILABLE* and are connected to the parallel nets, as shown in Figure 17.

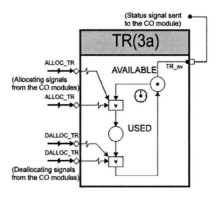

Figure 16. Model of resource Type 3 with the exclusively used device.

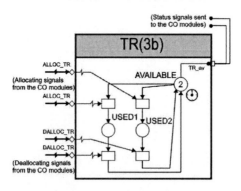

Figure 17. Model of resource Type 3 with the non-exclusively used device.

Resource Type 2

This resource is most often found in the plants that belong to Type 2. Main vessels and product storage tanks bearing the property of an internally buffer are such examples. The resource is charged by a particular control operation and is then discharged by a succeeding control operation. The model representing the simplest behavior of the resource type can be seen in Figure 18. An empty buffer, represented by place *EMPTY,* forwards a discrete status signal to the condition output *TR_av* which afterwards is evaluated by the requesting control operation. Once the buffer is ready (empty), the control operation charges the device via signal *START_CH* that moves the token from place *EMPTY* to place *CHARGE* until the charging is completed. Then, the token resides in place *FULL* to indicate a full state of the buffer. When the device is no longer needed and must be soon discharged, the discharging signal is sent by the succeeding control operation to move the token from place *FULL* to place *D_CHARGE* and to wait until the discharging process is finished. Afterward, the device returns available (empty) for upcoming processes.

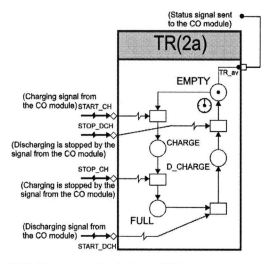

Figure 18. Model of resource Type 2 with one (dis)charging connection.

In some cases, resources Type 2 may be not used individually, yet they are exclusively shared by other parallel process units. As for example, the Biodiesel storage tank (see TR7 in column Plant A or TR8 in column Plant B) is used exclusively for storing each Biodiesel product produced by two plants. Based on the assumptions of the unlimited capacity of the final storage tank (very large), as well as the exclusive access by more than one production line, the model which meets the given specifications can be seen in Figure 19.

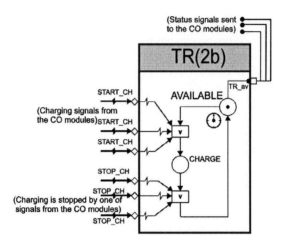

Figure 19. Model of resource Type 2 with exclusively used connection points.

In contrast to the previous assumption that presupposes the infinite capacity of resource Type 2, the following assumes a defined capacity. In this context, either the charging connection or the discharging connection can be used exclusively as well as non-exclusively depending on the existing limitations in the equipment. As shown in Figure 20, the maximum capacity of the main resource is denoted by the initial marking m occupying place *CAP_AVAILABLE*. Each charging process consumes n tokens which are taken from place *CAP_AVAILABLE* and then put on place *CAP_USED*. As long as the resource capacity is adequate (m>=n), the charging operation is permitted to be performed. The buffer is then emptied by the discharging operation which returns the buffer capacity as before (n). In some cases, the discharging operation is not controlled by the internal operation of the plant. To simulate the external unit, a time variable τ is added on the output arc of place *D_CHARGE*. Concerning on the restriction in the use of both the charging connection and the discharging connection, the places *CON_C_AV* and *CON_D_AV* are added as the status provider of the available connections. The daily tank in the second plant (TR7 in column Plant B in Table 14) is an example of this resource type.

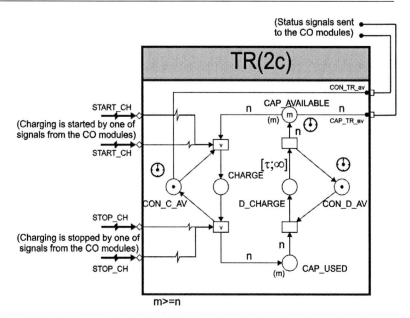

Figure 20. Model of resource Type 2 with a defined capacity and exclusively used connection points.

The following resource constitutes a variance of the previously described type. Common elements used in the model are the same with the previous one (Figure 20) unless the connection points are accessed non-exclusively by their clients. As presented in Figure 21, more than one the charging unit is set by the model to represent the non-exclusive access of the device. As long as the required capacity is available (m>=n), the charging of the resource can be performed unrestrictively. The secondary Product Tank (see TR8 in column Plant A in Table 14) in the first plant is belonging to this resource type.

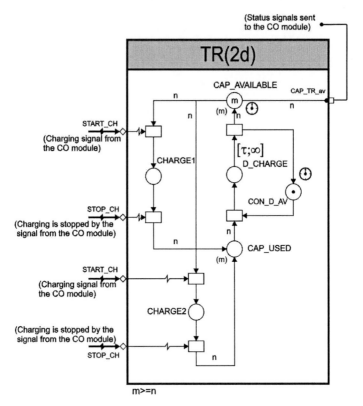

Figure 21. Model of resource Type 2 with a defined capacity and non-exclusively charging connections.

3.2.5 Basic Function Sequence

The software representative of a control operation (CO) is called basic function sequence (BFS). A basic function sequence describes the structural arrangement of the plant's technical functionalities (basic functions), underlying a particular process unit. The algorithm is realized by the sequential function chart in the basic function sequence module where each already executed basic function will send the feedback signal to the basic function sequence module, to move the current state to the next state (the upcoming execution), and so forth until all sequential tasks are finished.

To describe the relation among the underlying basic functions, there are three configuration options. The branches can be executed exclusively (OR branch) or commonly (AND branch) or even in the combination of both. The fact shows that the serial connection is more common than the parallel connection in basic function sequences. Clearly, Figure 22 displays the three proposed models.

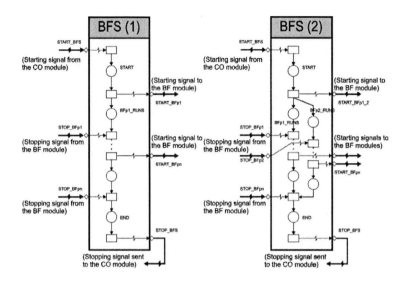

(a) Serial Connection   (b) Parallel Connection

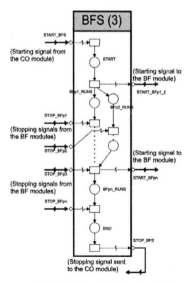

(C) Combination of serial and parallel connections

Figure 22. Model of the Basic Function Sequence.

3.2.6 Basic Functions

As introduced before, a basic function realizes the elementary level of operating a batch process. In this implementation, the defined basic function models are classifiable into three application types.

The model sketched in Figure 23 is the main basic function type dedicated for single process functionalities, meaning that the process step runs independently for τ time units and interruptions during the ongoing operation are not allowed. The executing order is accepted from a basic function sequence module and after the step runs, the stopping signal is fed back to the basic function sequence module.

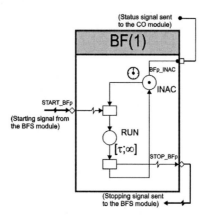

Figure 23. Model 1. of Basic Function.

In few cases, a basic function is not running independently, otherwise depending on the direct control action of another basic function. Hence, the processing time is not fixed, otherwise the end of the execution is controlled by the event signal connected to the other basic function module. As an implication, the time annotated at the arc, representing the process activity, is omitted. The functionality is then replaced by the event arc connected to the other module, as shown in Figure 24.

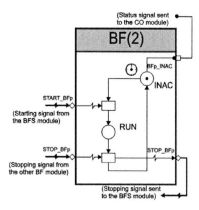

Figure 24. Model 2. of Basic Function.

Not much different from the model of the two basic function types above, the third type is dedicated for identical processes that run in different locations. As for example, the washing processes can be performed independently either in vessel *V1.n* or in vessel *V2.n*. Figure 25 models such specifications. As can be seen, two parallel nets inside of the model represent the two identical processes, which are applied to two independent, different locations.

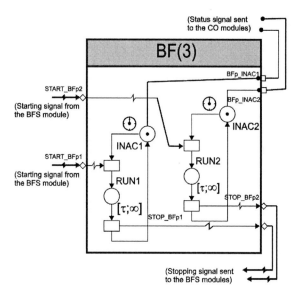

Figure 25. Model 3. of Basic Function.

The real processes that would be mapped into the basic function modules are summarized in Table 15. Basically, the process characterizing both plants is similar, unless the second plant operates more washing processes than the first plant. Therefore, the basic functions implemented to the second plant are more than the first plant, as described in Table 15.

The closed-loop model of the implemented modules of both plants is completely described in Appendix A1 (A1.1 – A1.4). Before analyzing the global model further on, the manufacturing requirements must be defined into formal specifications, as will be discussed in the next chapter.

Table 15. The implemented basic functions.

Basic Functions	Plant A	Plant B
BF1.n	Feeding the raw materials into V1.n for 80 tu.	Evacuation of the raw materials from the metering tanks into V3.n for 20 tu.
BF2.n	Heating V1.n for 74 tu.	Heating V3.n for 60 tu.
BF3.n	Stirring V1.n for 164 tu.	Stirring V3.n for 180 tu.
BF4.n	Keeping the temperature for 90 tu.	Keeping the temperature for 120 tu.
BF5.n	Cooling down V1.n for 51 tu.	Cooling down V3.n for 75 tu.
BF6.n	Discharging SP from V1.n into the SPT for 65 tu.	Discharging SP from V3.n into the tote tanks for 45 tu.
BF7.n	Adding D for Washing 1 for 15 tu.	Adding D for Washing 1 for 7 tu.
BF8.n	Stirring during Washing 1 for 15 tu.	Stirring during Washing 1 for 7 tu.
BF9.n	The process settles for 30 tu (Settling 1).	The process settles for 45 tu (Settling 1).
BF10.n	Discharging D from the vessel for 47 tu.	Discharging D from the vessel for 45 tu.
BF11.n	Adding D for Washing 2 for 11 tu.	Adding D for Washing 2 for 7 tu.
BF12.n	Stirring during Washing 2 for 11 tu.	Stirring during Washing 2 for 7 tu
BF13.n	The process settles for 30 tu (Settling 2).	The process settles for 30 tu (Settling 2).
BF14.n	Discharging D from the vessel for 15 tu.	Discharging D from the vessel for 20 tu.
BF15.n	V1.n is emptied by transferring the crude BD into V2.n for 36 tu.	V3.n is emptied by transferring the crude BD into V4.n for 20 tu.
BF16.n	Heating V2.n for 109 tu.	Adding D for Washing 3 for 7 tu.
BF17.n	Stirring V2.n for some time*.	Stirring during Washing 2 for 7 tu.

BF18.n	Keeping the temperature and drying for 177 tu.	The process settles for 30 tu (Settling 2).
BF19.n	Cooling down V2.n and stopping the stirring for 49 tu.	Discharging D from the vessel for 20 tu.
BF20.n	Storing BD for 102 tu.	Heating V4.n for 90 tu.
BF21.n	-	Stirring V4.n for some time*.
BF22.n	-	Keeping the temperature and drying for 90 tu.
BF23.n	-	Cooling down V4.n and stopping the stirring for 120 tu.
BF24.n	-	Storing BD into the daily tank (DT) for 30 tu.

SP :   secondary product, SPT: secondary product tank, BD: Biodiesel, tu: time units
* :   For BF 17 of Plant A, the duration depends on BF16.n plus BF18.n while for BF21 of Plant B, the duration depends on BF20.n plus BF22.n.
n :   the line number

4. Formal Specifications of the Batch-Plants

As it is pointed out in the previous chapter, control strategy as an output of analysis of a closed-loop model of plants requires plant specifications which are defined formally. In the presented case, formalization on the plant specifications encompasses two things, i.e. waiting periods of process devices corresponding to the optimal objective function and dangerous idle time related to the safety goal. Since the production sequences are performed cyclically, the plant specifications must be defined in context of the cyclic plant behavior. Therefore, at the beginning, the study about cyclic behavior is introduced.

4.1 Cyclic Behavior

Consider a cyclic model that is built based on an assumption of a stable demand of market with constant rate, as sketched in Figure 26. The model consists of three modules describing two process lines coupled by a conflict resource. The unlimited cyclic behavior can be obviously viewed at the end transition connected to the initial place of each module, which means that the set of the operations runs infinitely. The waiting places W_OP and W_OQ added in the model serve as buffer places to avoid deadlocks when the resource requested by one of the process lines is unavailable on time.

In practice, plant operations are in many cases limited (finite). The cyclic behavior of the production system is usually limited by the ordered batches denoted by a fixed number. The batch number that corresponds to the market demand and material supply determines how many times products must be produced within a time period. The model that meets the limited cyclic behavior can be seen in Figure 27. In the model, the place with the initial marking R denotes the process requests. When the process runs and afterward is repeated cyclically, one by one token is moved from the place R to the succeeding place (R) by firing transition tr until the place R is empty, representing that all requests are already fulfilled. Once the manufacturing process ends, the corresponding tokens must then be removed from the model by firing the transitions P_END and Q_END.

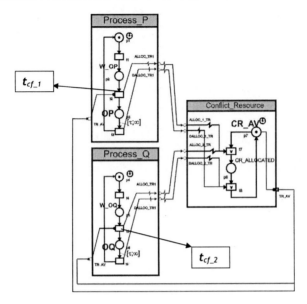

⊕ : Local time (denoting the token age), t_{cf_n} : conflict transition

Figure 26. TNCES model of a simple batch process.

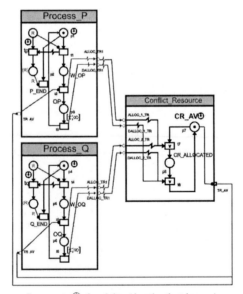

R : request, ⊕ : Local time (denoting the token age)

Figure 27. Limited cyclic model

As a consequence of such cyclic models, the corresponding nodes (representing the marking positions and also the times), generated by the dynamic graph, form cycle patterns too. Formalization on the cyclic paths is described in detail as below.

Definition 4.1 (Path of nodes in the dynamic graph):

A path of the nodes (notated with p) and the firing sequences which initiates from startup phases and then forms cycles can be defined as follows:

$$p = \left(\begin{array}{l} (\overline{z_0}, \overline{\omega_0}), \dots, (\overline{z_{i-1}}, \overline{\omega_{i-1}}), (\overline{z_i}, \overline{\omega_i}), \dots \\ \qquad \dots, (\overline{z_k}, \overline{\omega_k}) \end{array} \middle| \begin{array}{l} \overline{z_{i-1}}[\overline{\omega_{i-1}}\rangle\overline{z_i} \wedge \overline{z_k}[\overline{\omega_k}\rangle\overline{z_i} \\ \wedge \forall i,j \in \{0..k\} \\ (i \neq j \rightarrow \overline{z_i} \neq \overline{z_j}) \end{array} \right)$$

where the set of $(\overline{z_0}, \dots, \overline{z_{i-1}}, \overline{z_i}, \dots, \overline{z_k})$ refers to all nodes that maps the marking positions and also the times. The set of $(\overline{\omega_0}, \dots, \overline{\omega_{i-1}}, \overline{\omega_i}, \dots, \overline{\omega_k})$ is the set of firing sequences and the abbreviation $\overline{z_{i-1}}[\overline{\omega_{i-1}}\rangle\overline{z_i}$ denotes that state $\overline{z_i}$ can be reached from state $\overline{z_{i-1}}$ by the firing step $\overline{\omega_{i-1}}$. Additionally, the abbreviation $\overline{z_k}[\overline{\omega_k}\rangle\overline{z_i}$ denotes the path going back to its initial nodes by the firing step $\overline{\omega_k}$ (cycle). The readers that are interested to know more about the definition of the nodes and the firing sequences are referred to [27, 40].

Moreover, supposing that the notations $\overline{z}(b) = (\overline{z_0}, \dots, \overline{z_{i-1}})$ are the set of the nodes of the startup phases, $\overline{z}(c) = (\overline{z_i}, \dots, \overline{z_k})$ is the set of the nodes of the cycles while $\overline{z}(p)$ is the set of the nodes of the selected path p. The complete relation between $\overline{z}(b)$, $\overline{z}(c)$ and $\overline{z}(p)$ can be therefore written as follows:

$$\overline{z}(p) = \left((\overline{z}(b)), (\overline{z}(c))^\gamma \right)$$

Here each node of $\overline{z}(p)$ contains the marking position $\overline{m}$ and also the local time $\overline{l}$, while γ is the number of cycles.

Afterwards, the optimal control strategy can be synthesized by selecting one path based on a certain criterion. In this case the used optimal criterion is the waiting periods of process devices, as described in the following.

4.2   Waiting Periods

In batch processes, waiting periods occurring in main process devices such as tanks or vessels can be induced by three specific conditions, i.e.:

[1]   Discontinuity of process steps in main process devices due to the unavailability of conflict resources when the process line tries to access them.

[2]   The idle time of main process devices caused by the unused conditions, e.g. empty tanks during startup or empty tanks after finishing their tasks.

[3]   The idle time of main process devices caused by the queue (waiting) for the upcoming process transfer (between tanks).

These points will be formally defined in the following.

Definition 4.2 (Halt times) :

In models, to avoid deadlocks due to conflict resources, the waiting places with the untimed arcs are added. The waiting places will hold tokens of the process steps when conflicts in accessing the resources happen. By computing how long the tokens are trapped in the waiting places, the duration of the discontinuity of process steps can be determined. The discontinued times are called as *halt time*. Figure 26 can be used to describe a related illustration. The model represents two processes, i.e. process P and process Q coupled by a conflict resource. When the resource is being allocated by one of the process lines, the other process line will be discontinued by residing the token in one of the waiting places i.e. W_OP or W_OQ until the needed resource is reavailable. The delay time of the retained process is the firing time of the appropriate conflict transition, e.g. t_{cf_n}, that depends on the duration of the local time of the marked waiting places (W_OP or W_OQ).

By assuming that in the selected trajectory $\overline{z}(p)$ there are β waiting nodes containing the halt times and taking into account the process behavior from startup until cycle, the accumulation of all the nodes denoted by $\overline{z_{ht}}$ can be expressed in the following trajectory set:

$$\overline{z_{ht}} = \left(\left(\overline{z_{h1}}, \dots, \overline{z_{h(\alpha-1)}} \right), \left(\overline{z_{h\alpha}}, \dots, \overline{z_{h\beta}} \right)^{\gamma} \right)$$

where $\overline{z_{ht}} \in \overline{z}(p)$ holds. The set $\left(\overline{z_{h1}}, \dots, \overline{z_{h(\alpha-1)}} \right)$ is the startup sequence of the nodes of the waiting places and the set $\left(\overline{z_{h\alpha}}, \dots, \overline{z_{h\beta}} \right)^{\gamma}$ is the sequence of the nodes of the waiting places within a cycle. The halt times denoted by D_{ht} are then computed by summing all the duration of the local time of the waiting places in $\overline{z_{ht}}$.

$$D_{ht} = \sum_{\delta=1...(\alpha-1)} \left(\overline{l_{h\delta}}\right) + \left(\sum_{\delta=\alpha...\beta} \left(\overline{l_{h\delta}}\right)\right) * \gamma$$

Definition 4.3 (Unused times) :

In general, unused conditions of main process devices can be illustrated by means of a following example. Suppose that a single production line consists of two main vessels and is equipped with one metering tank. The main devices are operating sequentially and continually. By assuming that the occupation time of the first vessel is longer than the second vessel, there remain the unused times in the second vessel, caused by the tasks in the second vessel, which are accomplished earlier than in the first vessel. Additionally, the idle vessels can also occur during the startup time, for example, as a time consequence of the initial filling of metering tanks.

The model representing the above illustration can be seen in Figure 28. It consists of four modules i.e. *Process_W* representing the whole manufacturing process, *MR_Vessel_1* and *MR_Vessel_2* modeling the behavior of the two vessels, and the last module *Met_Tank* describing the behavior of the metering tank. The unused times of the vessels are easily determined by computing the duration of the local time of the marking places *MR_1_IDLE* and *MR_2_IDLE*, denoting the current idle status of the appropriate process devices. By considering that in the selected trajectory $\overline{z}(p)$ the idle condition for all the process devices occurs in Ψ times ranging from startup to a cycle, then the sequence of the idle states is:

$$\overline{z_{un}} = \left(\left(\overline{z_{u1}}, ..., \overline{z_{u(\Phi-1)}}\right), \left(\overline{z_{u\Phi}}, ..., \overline{z_{u\Psi}}\right)^{\gamma}\right)$$

with $\overline{z_{un}} \in \overline{z}(p)$. The set $\left(\overline{z_{u1}}, ..., \overline{z_{u(\Phi-1)}}\right)$ is the startup sequence of the nodes of the idle nodes and $\left(\overline{z_{u\Phi}}, ..., \overline{z_{u\Psi}}\right)^{\gamma}$ is the sequence of the cyclic nodes of the idle nodes. The summing of the local times of the path $\overline{z_{un}}$ describes the totally unused times (D_{un}) i.e.:

$$D_{un} = \sum_{x=1...(\Phi-1)} \left(\overline{l_{ux}}\right) + \left(\sum_{x=\Phi...\Psi} \left(\overline{l_{ux}}\right)\right) * \gamma$$

From this formulation, the first term represents the unused times during the startup while the second term defines the unused times within a cycle.

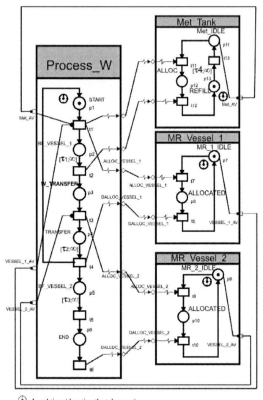

⊕ : Local time (denoting the token age)

Figure 28. Model of a process with two vessels and a metering tank.

Definition 4.4 (Queuing times) :

In contrast to the previous assumption (in def. 4.3), here one considers the occupation time of the second vessel being longer than the first vessel. By doing so, the process transfer from the first vessel to the second vessel must wait until all the ongoing tasks in the second vessel are completed. The waiting time is called as *queuing time*. The *queuing* condition is modelled by an intermediate place, inserted prior to the place denoting the process transfer, which has a function of holding the token of the sequence when the process transfer must wait for some time until the subsequent device is available. As for example, in Figure 28, the intermediate place is marked by W_TRANSFER. In general, in the dynamic graph, a number of σ the nodes of the waiting places, which are part of the selected trajectory $\overline{z}(p)$, is defined by the following trajectory equation:

$$\overline{z_{qt}} = \left(\left(\overline{z_{q1}}, \dots, \overline{z_{q(\varphi-1)}} \right), \left(\overline{z_{q\varphi}}, \dots, \overline{z_{q\sigma}} \right)^{\gamma} \right)$$

where the set $\left(\overline{z_{q1}}, \dots, \overline{z_{q(\varphi-1)}} \right)$ is the startup sequence of the queuing nodes and $\left(\overline{z_{q\varphi}}, \dots, \overline{z_{q\sigma}} \right)^{\gamma}$ is the cycle of the queuing states. The totally queuing time (D_{qt}) is the duration of the local time of the waiting places in $\overline{z_{qt}}$ which is counted from the startup phase up to the entrance of the cycle as expressed in the following:

$$D_{qt} = \sum_{\theta=1\dots(\varphi-1)} \left(\overline{l_{q\theta}} \right) + \left(\sum_{\theta=\varphi\dots\sigma} \left(\overline{l_{q\theta}} \right) \right) * \gamma$$

Definition 4.5 (Totally waiting periods) :

Finally, the totally waiting periods denoted by D_{wp} are defined by summing all the time durations D_{ht}, D_{un} and D_{qt}:

$$D_{wp} = D_{ht} + D_{un} + D_{qt}$$

Definition 4.6 (Optimal objective function and plant throughput) :

Related to these points, the desired behavior of the batch scheduling system is that the process devices at steady state conditions wait for as little as possible. Therefore, the optimal control strategy is the path that satisfies the minimal waiting period (WP) following the condition:

$$Obj = min \left(D_{wp} \right)$$

The plant performance is measured by means of the plant throughput. For each production cycle, it is computed based on the firing transitions t of the selected path $\overline{z}(p)$ that contributes to the output of product $O(c)$ as defined in reference [8] :

$$O(c) = \frac{\sum_{t \in T} B(t)occ(t,c)}{t_{Cycle}(c)}$$

where :

- ❖ $B : T \rightarrow \mathbb{N}$, B is defined as a value for the contribution of each node in the selected sequence to an objective function (for instance the output of product).
- ❖ $occ(t,c)$ denotes how often t fires in cycle c.
- ❖ $t_{Cycl}(c)$ denotes the cycle time of cycle c.

The value $B(t) = 1$ is given only for the nodes corresponding to the completion of each batch of a kind of product (which expresses the maximal value), otherwise $B(t) = 0$ for all other transitions.

4.3   Dangerous Idle Time

The second criterion used as the objective function for synthesizing the desired controller is related to the safety goal, i.e. dangerous idle time. The following defines the criterion formally.

Definition 4.7 (Dangerous states and the safe objective function) :

As for illustration, consider again the metering tank model in Figure 28 which is assumed to be containing a kind of dangerous material. The dangerous idle time is easily measured by computing the local time of the idle place *Met_IDLE* representing the idle condition of the metering tank when it is already refilled (filled) but not yet used. In the associated dynamic graph, the sequence $\overline{z_{dt}}$ that represents a number of ν dangerous states and then constituting part of the selected trajectory $\overline{z}(p)$ is formalized as follows:

$$\overline{z_{dt}} = \left(\left(\overline{z_{d1}}, \dots, \overline{z_{d(\pi-1)}} \right), \left(\overline{z_{d\pi}}, \dots, \overline{z_{dv}} \right)^\gamma \right)$$

where $\overline{z_{dt}} \in \overline{z}(p)$ holds, $\left(\overline{z_{d1}}, \ldots \overline{z_{d(\pi-1)}}\right)$ are the dangerous states during startup and $\left(\overline{z_{d\pi}}, \ldots, \overline{z_{dv}}\right)^{\gamma}$ are the dangerous states after entering the steady state. By summing all the local times of the corresponding nodes, the dangerous idle times D_{dt} can be computed by means of the following expression:

$$D_{dt} = \sum_{\partial=1\ldots(\pi-1)} \left(\overline{l_{d\partial}}\right) + \left(\sum_{\partial=\pi\ldots v} \left(\overline{l_{d\partial}}\right)\right) * \gamma$$

The first term represents the duration of the dangerous idle states during startup and the second term denotes the dangerous idle times within the cycle.

The objective function of the safety goal is formalized in the following:

$$Obj = min(D_{dt})$$

By analyzing a path which contains the minimum dangerous idle times, the control strategy which is free from the dangerous states (or at least the minimum dangerous states) is finally obtained.

These ideas will be applied to analyze the closed-loop model of the Biodiesel batch-plants as described in the following chapter.

5. Strategies for Tackling Complexity and Achieved Results

It has previously been predicted that complexity usually prevents analysis on large plants because of the size of the appropriate models. Hence, it is not surprising that the computations on the real plants have been terminated by the state explosion problems. Even for a simpler case, analysis on the global model of a single production line of one of the plants has shown that the dynamic graph computes a relatively large state space, i.e. about 2574 states with the complicated paths as shown in Figure 29. Thus, the optimal sequence of the case is difficult to be found. It is actually understandable, since the direct analysis approach (centralized approach) considers all reachable states including the unnecessary states, i.e. unoptimal states that exactly have the significant number in the computed dynamic graph. Although such a method can evaluate the whole model performance—including the unnecessary criteria in terms of optimization, it is only effective for small systems. Instead, for large systems, it would remain a problem in the state analysis.

Hence, to get rid of the burden of inherent complexity of the large system, the analysis approach must be oriented toward decomposition principles. One cannot directly analyze the complete plant model to find a globally optimal control strategy, but one can analyze smaller systems as a result of the decomposition method applied to the large system. By following the hierarchical, natural design of batch-plants itself, that consist of the process lines, the subsystems representing the single process lines can be analyzed step by step and finally are integrated into a global model to provide the desired behavior. However, this approach is better than a common strategy that relies on the experience approach. By the intuitive engineering approaches combined with the used formal modeling method, the proposed methods are proven to be able to cope with the complexity problems arising from the investigated real system, as it is summarized gradually in the following interrelated points.

[1] Complexity, which is caused by the model size of each single production line that is influenced by the recipe options.

[2] When the single production lines are coupled by the shared resources, the method for a determination of an optimal resource allocation for a commonly scheduling system of each plant suffers from the state explosion problem.

[3] Complexity of minimizing the dangerous idle times exists in the second plant, where the associated analysis must deal with the state explosion problem (the previous problem) and must also guarantee the optimal result.

[4]   Complexity of composing the two already optimized plant schedules into a globally
            scheduling system whereas the optimal behavior of each subsystem must be kept the
            same when unified.

The complexity problems above as well as the appropriate solutions will be explored
further in each following section.

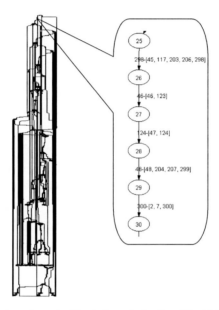

Figure 29. The dynamic graph of the single production line of Plant B (2574 states).

5.1   Complexity of a Single Production Line

Among two batch-plants, the most complicated case belongs to Plant B where the
analysis on each associated single production line has generated the dynamic graph with the
complicated paths (as shown in Figure 29). This is because the model is basically derived
from four basic recipe options that are unified flatly. As a consequence, all branches of the
reachable states of the dynamic graph grow exponentially following the number of the
composed options. Determination of the optimal trajectory of the states is therefore time-
consuming and even for typical cases, containing more equipment elements and more
operation options, such cases would be very difficult to be solved.

To overcome such complexity, instead of relying on the centralized approach which composes the whole recipe options into the global model, an alternative way is proposed. The strategy is to let the model of the process lines keep in its original single forms, namely the single model of the underlying basic recipe options. Each basic recipe option is classified based upon the location (vessels), where the underlying steps are performed. The duration of the steps, which for all basic recipe options is given by Table 3 up to Table 9 (see Chapter 2), is then summed independently to get the pure occupation time of each vessel. Table 16 describes the pure occupation time of each vessel along the production line in Plant A, while the same information for Plant B is given in Table 17. Based on the data, the pure occupation times are thereafter compared with each other to reselect recipe options that meet the criterion of the closest occupation time among consecutively used main devices along the production line. The selection has a purpose of minimizing the waiting times among the pair of the coupled recipe options. The selected recipes are then reconstructed into associated models for further analysis. By analyzing the smallest waiting period of the reconstructed models—under the assumption of the absence of conflicts of the use of the auxiliary resources along the single production line—the optimal sequence can be found. It is in line with the fact that conflicts only happen after the resources are coupled by more than one production line.

As for illustration, consider the fourth option of the control recipe (CR4) of Plant B as highlighted by the bold line in Table 17. Based on the information, CR4 implements its steps to vessel *V3.n* for the duration of 320 time units. The process is then continued in vessel *V4.n* for finishing the remaining steps (of CR4) for 541 time units. Until this point, one basic recipe option is already applied to the process line.

However, to maintain the continuity of the already started operation, once the process transfers from *V3.n* to *V4.n*, the concurrent process must be restarted in vessel V3.n. Here, one has four candidates of the basic recipe option which are applicable, namely reapplication of CR4 which means that the cycle behavior will be automatically created, or implementation of one of the other control recipe options, i.e. CR1, CR2 or CR3. The selected one (for V3.n) is the control recipe which has the occupation time as close as possible with the ongoing activity duration of vessel *V4.n*, which is under control of CR4 for 541 time units. In these respects, CR3 with the occupation time of 531 time units is obvious as the closest one, which is then selected as the pair of the coupled recipe option.

Table 16. The pure occupation times in Plant A.

Basic Recipes	OT	Remarks
CR1 (V1)	360	CR1 performed in V1, see Table 3
CR1 (V2)	585	CR1 performed in V2, see Table 3
CR2 (V1)	508	CR2 performed in V1, see Table 4
CR2 (V2)	437	CR2 performed in V2, see Table 4
CR3 (V1)	452	CR3 performed in V1, see Table 5
CR3 (V2)	493	CR3 performed in V2, see Table 5

OT: the pure occupation time in time units (tu).

Table 17. The pure occupation times in Plant B.

Basic Recipes	OT	Remarks
CR1 (V3)	474	CR1 performed in V3, see Table 6
CR1 (V4)	387	CR1 performed in V4, see Table 6
CR2 (V3)	417	CR2 performed in V3, see Table 7
CR2 (V4)	444	CR2 performed in V4, see Table 7
CR3 (V3)	531	CR3 performed in V3, see Table 8
CR3 (V4)	330	CR3 performed in V4, see Table 8
CR4 (V3)	320	CR4 performed in V3, see Table 9
CR4 (V4)	541	CR4 performed in V4, see Table 9

OT: the pure occupation time in time units (tu).

Afterwards, after the process transfers, the sequence is continued by accomplishing the remaining steps of CR3 in vessel *V4.n* for 330 time units. At the same time, to reutilize the vessel *V3.n* which is empty, the same idea mentioned above is reapplied. Here, CR4 with the closest occupation time, i.e. 320 time units, is once again selected to operate this vessel. By doing so, a cyclic behavior is eventually created. The flowchart visualizing the described example can be seen in Figure 30(d), while the flowcharts using the same method for all the decomposed models of the other recipe options are presented in Figure 30(a–c).

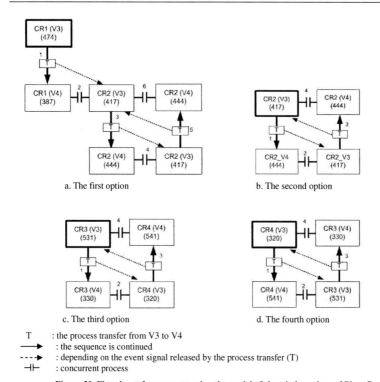

a. The first option   b. The second option

c. The third option   d. The fourth option

T   : the process transfer from V3 to V4
→   : the sequence is continued
---→   : depending on the event signal released by the process transfer (T)
-||-   : concurrent process

Figure 30. Flowcharts for reconstructing the model of the whole options of Plant B.

The detailed model satisfying the illustration of the flowchart (Figure 30 (d)) can be seen in Figure 31 and Figure 32. The model is originally reconstructed from the flat model of the single production line of Plant B (see Appendix A1.3) and is then modified by eliminating the main modules CR1 and CR2 including their following nets. If one takes a closer look into the modules CR4 and CR3 as sketched in Figure 32, CR4 has been redesigned to start the first execution in accordance with the initial condition mandated by the flowchart of the fourth option. Otherwise, the execution of CR3 must wait the event signal *CO6_4_END* that is released by CR4 once the initial process transfers to the next vessel. Although all steps of CR4 are already executed, the next cycle will not be restarted by itself, but depending on the event signal *CO6_3_END* that is controlled by CR3. By this modification, the model behaves consistent with the fourth option.

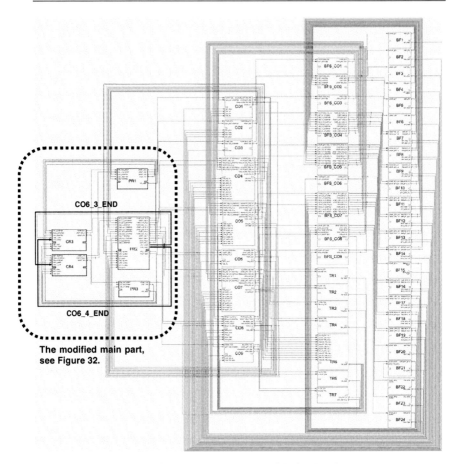

Figure 31. The reconstructed model of the fourth option.

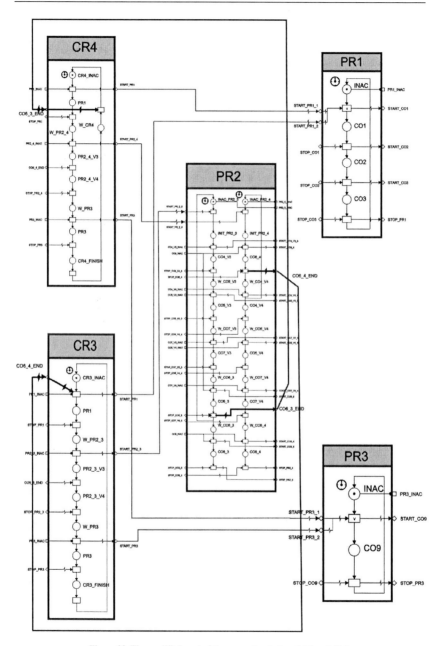

Figure 32. The modified part of the reconstructed model (partially).

The similar principle of these modeling ways is also applied to reconstruct the decomposed models of the other recipe options, where analysis on the whole models would be compared each other to find the optimal sequence. The limited reconstructed models eliminate the trajectories that fail to fulfill the requirement of the closest occupation times among the used main vessels (not optimal). Hence, states of the resulting dynamic graphs are reduced and the cycles are simplified. The whole dynamic graph of the four decomposed models of each single production line of Plant B generates only 1014 states containing 4 cycles, instead of the complicated paths of 2574 states contained by the flat model. The four paths then are analyzed by pointing out several variables, i.e. startup time (t_st_up), cycle time (tcyc), time of end of each batch (B.n), and also waiting periods (WP) of all paths. They are plotted into Table 18. The path which has the minimum waiting period can be found by comparing all the paths after reaching their cycles. The comparison point is fixed on the path which has the longest startup, to ensure that all the paths have entered their steady state conditions. The data in Table 18 show that the path p1 has the longest startup which starts its cycle in batch 2nd. Therefore all waiting periods of each path will be compared after batch 2nd (i.e. batch 2, 3,.. so on). For these reasons, the summing of the waiting periods (after the second batch) is denoted by a new symbol WP*. Analysis on the cyclic behavior of all the paths then shows that the path p4 (the fourth option) has the smallest waiting period i.e. WP* = 570+10(kb), kb=n-2, n>=2 (see rank 1 in Table 18), meaning that this one is the optimal solution for this case.

The associated Gantt-diagrams that compare the proposed optimal solution and the current operation of each single production line of Plant B are presented in Figure 33. They show the improved results. The proposed strategy p4 (the fourth option) uses the resources more efficiently. Although the unused time of vessel $V3.n$ keeps the same (i.e. 120 time units), the unused time of vessel $V4.n$ can be suppressed from 537 time units to 440 time units. Likewise the queuing time (in Figure 33 shown as "Q") of the process transfer from V3 to V4, it is successfully minimized, namely from 27 time units per batch to 10 time units per batch. Hence, the products will be manufactured faster. By cross-checking the data of the Table 18 (B.m), it is obvious that the path p4 has given the fastest production time of each production line. Further analysis until the tenth batch shows, that the production time can even save approximately 182 time units. Furthermore, the trend of this saved production time will increase linearly with respect to time in following batches.

Table 18. Data of startup, cycles, production time and waiting periods of one production line of Plant B.

Paths	tcyc	t_st_up	B1	B2	B3	B4	B5	B6	B7	B8	B9	B10	H	Un_V3	Un_V4	Q	WP	WP*	Rank
p1	464	1058	**1001**	**1495**	1959	2423	2887	3351	3815	4279	4743	5207	no	120	594+30	27k1	714, m=1 / 744+27k1, k1 = m-2, m>=2	744+27(kb)	4
p2/CA	464	564	**1001**	1465	1929	2393	2857	3321	3785	4249	4713	5177	no	120	537	27k2	657+27k2, k2=m-1,m>=1	684+27(kb)	3
p3	911	691	**1001**	**1562**	1912	2473	2823	3384	3734	4295	4645	5206	no	120	651	10k3	771+10k3, k3=m-1,m>=1	781+10(kb)	2
p4	911	467	**1001**	**1351**	1912	2262	2823	3173	3734	4084	4645	4995	no	120	440	10k4	560+10k4, k4=m-1,m>=1	570+10(kb)	1

Note:

bold text : the first cycle

t_st_up : start up time

tcyc : time of cycle

CA : current approach

Bm : end of batch.n or process of storing product.m has been finished (time units)

k is a constant that is a function of m, m is batch number (B.m)

* Waiting Period (WP*) has a function of kb, kb = m-2, m>=2

H: Halt, Un : Unused, Q: Queue

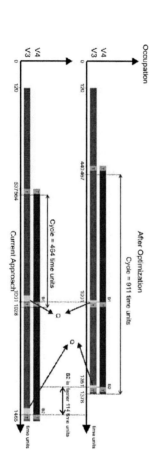

Figure 33. Occupation of main vessels in the single production line of Plant B.

Q : queue . T:transfer. B.n : End of batches with n: batch number

On the other hand, application of the same method, to simplify the huge model of each single production line of Plant A, results in the flowcharts used for reconstructing the decomposed model as shown in Figure 34. As for example, consider one of the options, i.e. the first option. At first, CR1 occupies vessel V1 for 360 time units and afterward controls vessel V2 for 585 time units. To reutilize vessel V1 (empty) during vessel V2 runs, the chosen basic recipe option is CR2 with the occupation time, i.e. 585 time units. The sequence of CR2 is then continued in vessel V2 for finishing the remaining steps for 437 time units. At the same time, to reutilize the vessel V1.n which is already empty, the same idea is reapplied. Here CR3 with the closest occupation time, i.e. 452 time units is selected to operate this vessel, and continuing for the following process until a cyclic behavior is at the end created.

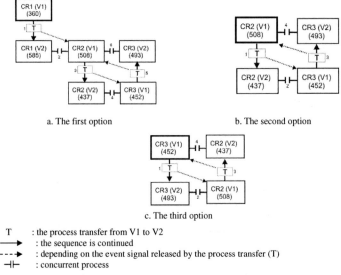

a. The first option   b. The second option

c. The third option

T   : the process transfer from V1 to V2
———▶   : the sequence is continued
- - -▶   : depending on the event signal released by the process transfer (T)
—||—   : concurrent process

Figure 34. Flowcharts for reconstructing the model of the whole options of Plant A.

All the reconstructed models generate about 708 states with 3 trajectories of the firing sequences, which are denoted by p1, p2 and p3. The performance evaluation on the paths is shown in Table 19 where the path p1 (the first option, see Figure 34(a)) is the optimal one. By mapping the optimal sequence into a Gantt chart and comparing it with the current operational approach as shown in Figure 35, one can then see how the performance of the plant changes. The proposed strategy p1 occupies the main vessels more efficiently such that after the second batch and so on it decreases the inactive phases of the vessel V2 from 71 time units per batch to 15 time units per batch (see the unused spaces between the vessel occupation). Although the queue of the process transfer between the consecutively used devices (in Figure 35 shown as "Q") still exists, i.e. about 77 time units, it only occurs during the startup condition. After reaching the steady state, no queuing times are necessary. Hence, the production behavior will be accelerated. By cross-checking the end time of each batch (B.m) for all paths in the Table 19, it can be clearly seen that the path p1 has given the fastest production time of the plant. Further analysis until the batch 10 shows then, that the production time which can be saved reaches approximately 224 time units. Furthermore, the trend of this saved production time will increase linearly with respect to time in the following batches.

Table 19. Data of startup, cycles, production time and waiting periods of one production line of Plant A.

Paths	tcyc	t_st_up	B1	B2	B3	B4	B5	B6	B7	B8	B9	B10	H	Un V1	Un V2	Q	WP	WP*	Rank
p1	1032	981	**981**	**1454**	1998	2486	3030	3518	4062	4550	5094	5582		360+15(k1)	77	360, m =1	360+15(k1), k1=m-2,n>=2	452+15(ka)	1
p2	1032	493	**981**	**1525**	2013	2557	3045	3589	4077	4621	5109	5653	No	508+15(k2)			437+15(k1), k1=m-2,n>=2; 508+15(k2), k2 = m-1, m>=1	538+15(ka)	3
p3	1032	440	**981**	**1469**	2013	2501	3045	3533	4077	4565	5109	5597	No	452+15(k3)			452+15(k3),k3=(m-1),m>=1	482+15(ka)	2
CA	544	981	**981**	**1454**	1998	2542	3086	3630	4174	4718	5262	5806	No	360+71(k4)	77	360, m =1	437+71(k4), k4=m-2,m>=2	508+71(ka)	4

Note:

bold text: the first cycle

t_st_up : start up time

tcyc : time of cycle

CA : current approach

B.m : end of batch.m or process of storing product.m has been finished (time units)

ka is a constant that is a function of m, m is batch number (B.m)

* Waiting Period (WP*) has a function of ka, ka = m-3, m>=3

H: Halt, Un: Unused, Q: Queue

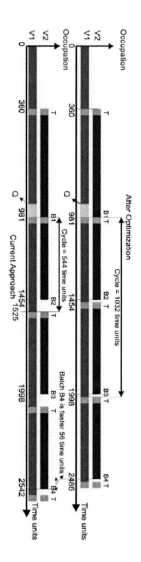

Figure 35. Occupation of main vessels in the single production line of Plant A.

Q : queue , T:transfer, B.n : End of batches with n: batch number

After the single production lines are optimized individually, the further step is to analyze the optimal control strategy for each plant that is derived from the model of the single production lines interconnected with the model of the shared resources. Due to the coupling, there are two possibilities of conflicts. First, only temporary conflicts are detected. The existing deviations should be able to be resolved by the control strategy only in the startup sequence or in other words, the expected control strategy will result in the permanent plant behavior which is free from the conflicts. By doing so, at the end, the optimal sequence of each plant exactly represents the optimal sequence of the contained single production lines and this will be proven in the next section. Second, permanent conflicts arise, for example, as a consequence of the restriction of the process design itself. The available time slot of the use of the limited resources cannot satisfy the whole requests of the process lines permanently, so that the waiting times (halt times) are finally required. However, since the investigated case does not indicate the existence of the permanent conflicts, those will not be further considered in this thesis. The following section focuses on the crucial problem accompanying the temporary conflicts, and a suitable strategy to cope with it.

5.2 Production Lines of Each Plant Coupled by Shared Resources

As it is known, each plant consists of two single production lines coupled by the exclusively shared resources (see Chapter 2). Hence, there is exactly a substantial need to fuse the single production schedules into a common scheduling system of each plant. Unfortunately, analysis on the global model of each plant shows, that determination of the resource allocation for the coupled process lines was stuck in the state explosion problem, terminating computation of the dynamic graph at more than 100.000 states. Supposing that the computation can be accomplished, the state number is too large for further analysis, so the optimal control strategy is very hard to be synthesized. Therefore, there is a need of an alternative approach that can simplify the complexity.

The proposed strategy is motivated by reference [9] and is strongly related to the prior work of the author in [6] which is called with the modular strategy. The novel approach uses modularity principles which decompose a large system into small systems (subsystems) and analyzes them stage by stage. A transformation by reconstructing the firing transitions of the optimal paths of the isolated subsystems into a purely sequential model called as "transformed model" is then done. The transformed model describes the logical and temporal order in which the different processes in the different process lines are started and executed. It includes the startup phase as well as the cyclic, stationary behavior of the modeled subsystems. In other words, the transformed model contains optimal paths of subsystems recoupled by the shared resources. It is proven by simulation that the behavior of the transformed model exactly represents the original optimal path of the original model.

However, more effort must be given for constructing the transformed model of large plants. Moreover, the behavior of the subsystems forms the startup sequences as well as the cycles with long time constants, so that the behavior leads to the size of the transformed model bigger than the original one. In fact, the transformation itself is intended to reduce the behavior of the original model. Therefore, in the proposed method, the purely transformed model is no longer used. As a replacement, the reduced behavior of the original model is adequately represented by modifying the original model itself, particularly on the net of the resource allocation, so as the original model can be "forced" to contain only the desired behavior, i.e. optimal sequence of the contained subsystems. This strategy is called the refinement method. By this way, efforts to reconstruct the optimal sequence of the subsystems of the large plants can be minimized. Clearly, the method is described in the following.

In general, the proposed method is initiated by decomposing the large system into smaller ones (subsystems) and analyzing the optimal path of each subsystem separately. The decomposition criterion corresponds to the plant model that is partitioned following the natural forms of the plant process itself, which consists of process lines (modules) bounded by the shared resources. Although the subsystems are not referring to the most basic modules that use sensors and actuators as the "smallest criterion", however, the proposed boundary has been proven to be able to simplify the complexity caused by the state explosion problem. The result of the component analysis is afterward used as a basis for refining the behavior of the original model. Thus, the refined original model only contains the desired behavior, i.e. the optimal subsystems which are eventually recoupled by means of the shared resources. Since the refinement strategy reduces the original behavior, the resulting states in the dynamic graph, by itself, are decreased drastically.

By considering the application scope that is intended to batch-plants with identical production lines (identical behaviors), and also by assuming that no conflicts will change the optimal behavior of the whole system permanently, the decomposition technique can be expected to give an optimal result globally. To convince the optimal criterion, the final result will be checked against the optimal parameters of each contained subsystem. As long as the waiting periods keep the same after the subsystems are composed (no permanent conflicts are detected in a cycle), as indicated by the constant cycle time, the final result can be ensured to be optimal, accordingly.

Systematically, the proposed idea is realized by the following steps:

Step 1: decompose the complete model into several subsystems following the plant process lines. Each subsystem can consist of one or more process lines depending on the intuitive analysis of "the maximal size" of the contained process lines. As for example, a system consists of six identical process lines. Assuming that a subsystem is still analyzable (by means of the dynamic graph) if consisting of three process lines maximal or less, then the system should be decomposed at least into two subsystems, where every subsystem consists of three process lines (maximal).

Step 2: analyze the dynamic graph of each subsystem to find the optimal path of the subsystems. Translate the optimal paths into the sequence of the state transitions. Focus on the firing transitions corresponding to the resource allocation. Based on the transitions, investigate how the optimal path of the subsystems manages the shared resources in a priority sequence.

Step 3: add the temporary subnets on the original model with the aim of controlling the signal flow of the resource allocation, so that it is exactly matching with the sequence priority of the optimal resource allocation. By this way, the behavior of the original model can be reduced since it only contains the desired behaviors, i.e. the optimal subsystems. The modified original model is called "the refined model".

Step 4: analyze the dynamic graph of the refined model. Based on the selected criterion, find the optimal path and map it into the firing transitions. By omitting the artificial transitions which correspond to the temporal subnets, the solution that exactly represents the optimal path of the original model is finally found.

The refinement strategy can be seen as a method which has a filter function with respect to unnecessary states. Step 2 and Step 3 have obviously reduced the number of states of the original dynamic graph since the refined model only contains the optimal states of the subsystems, while the unnecessary states (the unoptimal states) are not included at all.

To demonstrate the idea, Figure 36 presents a model of a simplified batch-plant. The model represents five identical process lines coupled with a conflict resource. The process lines are modeled by the modules LINE_n (with n as the number of the process lines) depicted in Figure 36(a), while Figure 36(b) models the content of the appropriate subnets. Operations along each process line are denoted by BF1_L.n and BF2_L.n where the steps require the time durations, i.e. 2 time units and 30 time units, respectively. Resource allocation is realized by the step BF1_Ln with triggering the allocating signal to the resource module modeled in Figure 36(c). To optimize the system, a shortest cycle is considered as the optimal criterion. This means that one must select a path of state transitions containing the shortest cycle for all the process lines.

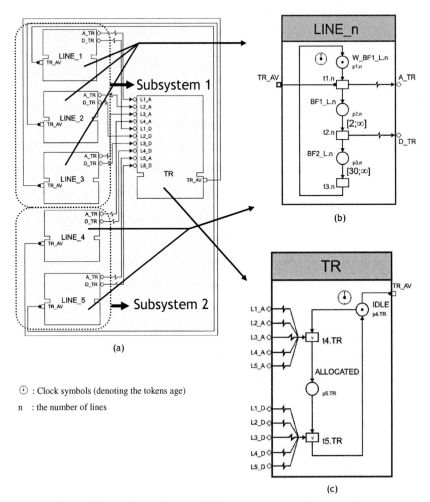

(a)

⊙ : Clock symbols (denoting the tokens age)

n : the number of lines

Figure 36. Model of the batch-plant with five production lines and one shared resource.

Application of the proposed strategy above to the model example would be described as follows.

Step 1: decompose the whole model into several subsystems. For this case, Subsystem 1 is Line 1, Line 2 and Line 3, while Subsystem 2 is Line 4 and Line 5 (see Figure 36(a)).

Step 2: For Subsystem 1, based on the dynamic graph analysis, the shortest sequence of the transition nodes is as follows:

p_{subs-1} = {$\underline{t_{1.1}}$, $t_{4.TR}$}(0), {$t_{2.1}$, $t_{5.TR}$}(2), {$\underline{t_{1.2}}$, $t_{4.TR}$}(0), {$t_{2.2}$, $t_{5.TR}$}(2), [{$\underline{t_{1.3}}$, $t_{4.TR}$}(0), {$t_{2.3}$, $t_{5.TR}$}(2), {$t_{3.1}$}(26), {$\underline{t_{1.1}}$, $t_{4.TR}$}(0), {$t_{3.2},t_{2.1}$, $t_{5.TR}$}(2), {$\underline{t_{1.2}}$, $t_{4.TR}$}(0), {$t_{3.3},t_{2.2}$, $t_{5.TR}$}(2)]^m

while for Subsystem 2, the shortest sequence is:

p_{subs-2} ={$\underline{t_{1.4}}$, $t_{4.TR}$}(0), {$t_{2.4}$, $t_{5.TR}$}(2), [{$\underline{t_{1.5}}$, $t_{4.TR}$}(0), {$t_{2.5}$, $t_{5.TR}$}(2), {$t_{3.4}$}(28), {$\underline{t_{1.4}}$, $t_{4.TR}$}(0), {$t_{3.5},t_{2.4}$, $t_{5.TR}$}(2)]^m.

(The bold text is the cycle, m = the cycle number).

According to $p_{subs-1}(s)$, the resource allocation is started for Line_1 and then followed for Line_2 and finally given for Line_3 (see the underlined transitions $\underline{t_{1.1}}$, $\underline{t_{1.2}}$, $\underline{t_{1.3}}$). On the other hand, $p_{subs-2}(s)$ arranges the resource allocation respectively for Line_4 and Line_5 (see the underlined transitions $\underline{t_{1.4}}$, $\underline{t_{1.5}}$).

Step 3: in this step, the model refinement is done. In order to force the original behavior to be reduced only containing the optimal sequences, i.e. $p_{subs-1}(s)$ all at once $p_{subs-2}(s)$, the original model is modified by adding the event signals *PR_α*, *PR_β* and *PR_λ* linked with the associated subnets to construct the priority sequence of the optimal resource allocation. The refined model is given in Figure 37 with the bold nets representing the desired priority sequence.

Step 4: Afterward the refined model is analyzed. The dynamic graph consists of 10 paths containing 10 identical cycles with 226 states. The corresponding shortest cycle has the cycle time of 40 time units and has the state transitions as follows:

p_{ref} ={$t_{1.1},t_{4.TR}$}(0), {$t_{2.1}$, $t_{5.TR}$, $t_α$}(2), {$t_{1.2}$, $t_{4.TR}$}(0), {$t_{2.2}$, $t_{5.TR}$, $t_β$}(2), {$t_{1.3}$, $t_{4.TR}$}(0), {$t_{2.3}$, $t_{5.TR}$}(2), {$t_{1.4}$, $t_{4.TR}$}(0), {$t_{2.4}$, $t_{5.TR}$, $t_λ$}(2) [{$t_{1.5},t_{4.TR}$}(0), {$t_{2.5}$, $t_{5.TR}$}(2), {$t_{3.1}$}(22), {$t_{1.1}$, $t_{4.TR}$}(0), {$t_{3.2}$, $t_{2.1}$, $t_{5.TR}$, $t_α$}(2), {$t_{1.2}$, $t_{4.TR}$}(0), {$t_{3.3}$, $t_{2.2}$, $t_{5.TR}$, $t_β$}(2), {$t_{1.3},t_{4.TR}$}(0), {$t_{3.4}$, $t_{2.3}$, $t_{5.TR}$}(2), {$t_{1.4}$, $t_{4.TR}$}(0), {$t_{3.5}$, $t_{2.4}$, $t_{5.TR}$, $t_λ$}(2)]^m.

(The bold text is the cycle, m = the cycle number).

By neglecting the artificial transitions $t_α$, $t_β$, $t_λ$ which are not representing the actual processes, the control solution for the system is finally obtained.

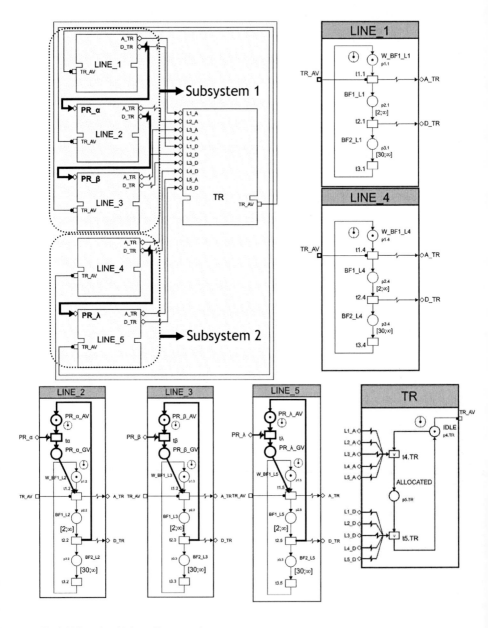

The bold lines: the added part (the new nets)

Figure 37. The refined model of the simplified batch-plant.

To examine the result of the proposed strategy, the global model of the plant is directly analyzed for comparison (the centralized approach). Actually, the proposed example (for this case) is simplified to make the comparison analysis easier. Nevertheless, the number of states of the resulting dynamic graph is big enough, namely around 2570 states with 120 identical cycles. The shortest sequence is:

p_{origin} = {$t_{1.1}$,$t_{4.TR}$}(0), {$t_{2.1}$, $t_{5.TR}$}(2), {$t_{1.2}$, $t_{4.TR}$}(0), {$t_{2.2}$, $t_{5.TR}$}(2), {$t_{1.3}$, $t_{4.TR}$}(0), {$t_{2.3}$, $t_{5.TR}$}(2), {$t_{1.4}$, $t_{4.TR}$}(0), {$t_{2.4}$, $t_{5.TR}$}(2) [{**$t_{1.5}$,$t_{4.TR}$**}**(0), {$t_{2.5}$, $t_{5.TR}$}(2), {$t_{3.1}$}(22), {$t_{1.1}$, $t_{4.TR}$}(0),** {**$t_{3.2}$,$t_{2.1}$,$t_{5.TR}$**}**(2), {$t_{1.2}$, $t_{4.TR}$}(0), {$t_{3.3}$,$t_{2.2}$,$t_{5.TR}$}(2), {$t_{1.3}$,$t_{4.TR}$}(0), {$t_{3.4}$,$t_{2.3}$, $t_{5.TR}$}(2), {$t_{1.4}$, $t_{4.TR}$}(0), {$t_{3.5}$, $t_{2.4}$, $t_{5.TR}$}(2)]**^m.

(The bold text is the cycle, m = the cycle number)

From the comparison of p_{ref} and p_{origin}, the shortest sequence of both approaches is principally identical. Therefore, it can be expected that this refinement strategy is not only giving a sub-optimal result, but even an optimal one. This method reduces the states drastically. The dynamic graph of the global model consists of 2570 states with 120 identical cycles, whereas the use of the refinement strategy only results in 226 states with 10 cycles. Therefore, this novel approach promises simplicity for analyzing large systems.

Implementation of the refinement strategy to the actual plants is described in the following. It will start with the first plant.

5.2.1 Analysis on Plant A

First, since the plant has two identical process lines, the global model must be partitioned into two subsystems, denoted by *SUBSYSTEM_1* and *SUBSYSTEM_2*, where each subsystem is each single production line itself.

Second, there is a need to analyze the optimal path of each subsystem. As it is described earlier, the waiting period of the process devices is used as the criterion to optimize each single production line. Besides, it has also been shown that the path p1 with the cycle time of 1032 time units is the optimal solution for each single production line of this plant (see Table 19).

Third, create the refined model based on the optimal sequence of the subsystems obtained by the previous step. The refined model of the appropriate plant is completely presented in Appendix A2.1. In the presented model, the module *SUBSYSTEM_1* stands for the optimal sequence of the first production line, whereas the module *SUBSYSTEM_2* denotes the optimal sequence of the second production line. The modules in the middle part interpret the behavior of the used resources.

Fourth, analyze the dynamic graph of the refined model for synthesizing the optimal control strategy. The refined model has drastically reduced the dynamic graph, i.e. only 2573 states with 2 cycles as shown in Figure 38. Compare it with the global model generating more than 100.000 states! One of the cycles is the desired optimal solution.

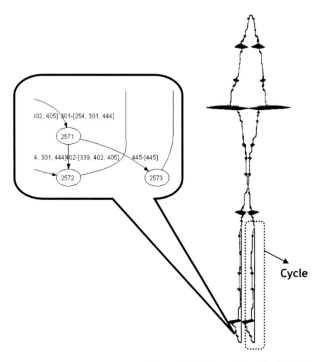

Figure 38. The dynamic graph of Plant A.

The control solution summarized and compared with the current approach is given in Table 20 where the cycle time of the optimized plant is 1032 time units, exactly the same with the production cycle of each individually optimized single production line (see Table 19). This means, that the optimal behavior of each single production line does not change at all after the process lines are unified into the common scheduling system. Likewise, if one take a closer look at the waiting period (WP*) on the plant scheduling, it keeps constant, i.e. 15 time units for each single production line under the cycle time of 1032 time units. In other words, the composed system is optimal.

Table 20. Optimal solution and current approach of Plant A.

Paths	tcyc	t_st_up	B1.1	B1.2	B2.1	B2.2	B3.1	B3.2	B4.1	B4.2	B5.1	B5.2	Un_V2
OA	1032	2103	984	1086	1457	1559	2001	2103	**2489**	**2592**	**3033**	**3136**	15
CA	544	1559	984	1086	1457	1559	**2001**	**2104**	2545	2648	3089	3192	71

Bm.n : end of batch.m (time units), n = the number of process line, bold text: the first cycle
Un_V2 = unused time (in time units) of vessel V2 in each cycle in each production line
t_st_up : start up time   CA : Current Approach
tcyc : cycle time   OA: Optimal Approach

By mapping the optimal states into the Gantt-diagram as depicted in Figure 39, the plant performance can be completely evaluated. The comparison between the proposed control strategy and the current approach of the plant provides the improved results. During the production runs, the discontinuity of the process steps caused by the unavailability of the conflict resources (in Figure 39 that is symbolized with H) occur at 0 time units, at 488 time units and at 959 time units where all the conflicts only arise during the startup phase. No permanent conflicts occur within the cycle. Likewise, with the queues (Q) for the process transfer from $V1.n$ to $V2.n$, they are only required during the process initialization. The proposed control strategy has also decreased inactive phases of the main process devices (unused) as indicated by the reduced blank spaces between the vessel utilization. Put differently, the main process devices can be occupied efficiently by the control strategy so that the production speed can be increased. As seen in Figure 39, the first cycle has successfully accelerated the batches 4th and 5th up to 56 time units. Until the tenth batch, the control strategy can speed up the production time of each process line up to 224 time units (or 448 time units for the both lines) and will increase linearly in the following batches (until the batch .20, the acceleration of the production speed of both lines can reach 1008 time units and so on).

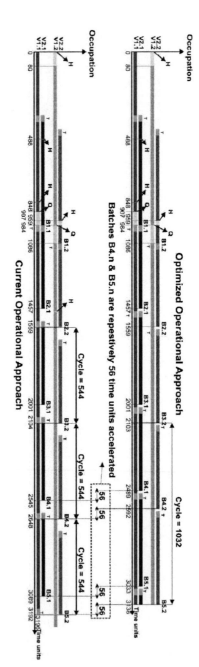

Vp.q: Vessel, T: Transfer, H: Halt, Q: Queue, Bm.n: End of batch

Figure 39. Performance evaluation of Plant A.

By converting the time acceleration to the plant throughput based on the equation of definition 4.7, the increase of the throughput is calculated about 0.0002 batches/time unit or 5.4% more productive than the current approach, as compared by Table 21. By considering the plant capacity of 28.5 MT/ batch, the production time in minute and also the data of benchmark prices of Biodiesel i.e. 1,013 USD/MT [39], the increased throughput per day ($\Delta O(c)$) is estimated about

$$\Delta O(c) = 0.0002 \; x \; 24 \; x \; 60 \; x \; 28.5 \frac{MT}{day}$$

$$\Delta O(c) = 8.208 \; MT/day$$

or equivalent to a value of 8,314 USD/day.

Table 21. Throughputs of Plant A.

Path	Throughput [Batches/Time unit]
OA	0.003676
CA	0.003876
Δ Throughput	0.0002 or ↑ 5.4%

5.2.2   Analysis on Plant B

Since the second plant is composed by two identical process lines, the corresponding global model is also divided into two subsystems. The first process line is denoted as subsystem 1, whereas the second one corresponds to subsystem 2. Later on, each subsystem is separately analyzed to find the optimal sequence. Based upon the earlier analysis results, the path Q_4 with the cycle time of 911 time units (see Table 18) is the optimal sequence for each single production line. The optimal sequence of the subsystems is reconstructed into the refined model for further analysis. One can see Appendix A2.2 for the more detailed model. The resulting dynamic graph shows that the refinement strategy successfully reduces the states significantly. As shown in Figure 40, the dynamic behavior of the refined model becomes much simpler, i.e. about 1563 states with 2 identical cycles, instead of the computation of the global model that results in more than 100.000 states. One of the cycles is the optimal control solution for this plant.

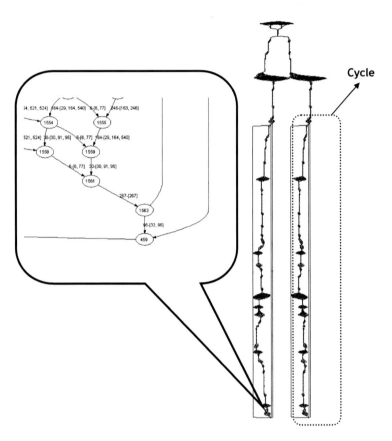

Figure 40. The dynamic graph of Plant B.

Further analysis on the optimal control strategy (OA) against its current approach (CA), as summarized in Table 22 and mapped into Figure 41, shows that the cycle time of the commonly scheduling system is calculated at 911 time units, precisely the same with the cycle time of the individually optimized single schedule as shown in Table 18. After the steady states, the waiting period is successfully suppressed from 27 time units (the current approach) to 10 time units (see Q in Figure 41 and in Table 22) per batch. No following conflicts happen within the cycle of the sequence, meaning that the commonly scheduling system of the plant remains optimal although the single production lines are coupled by the exclusively shared resources. By doing so, the products would be manufactured faster. It is measurable that the second batch (B2.n) can be accelerated approximately 114 time units earlier than its current approach and would be slightly increased in the following batches.

Table 22. Optimal solution and current approach of Plant B.

Paths	tcyc	t_st_up	B1.1	B1.2	B2.1	B2.2	B3.1	B3.2	B4.1	B4.2	B5.1	B5.2	Q
CA	464	704	**1001**	**1141**	1465	1605	1929	2069	2393	2533	2857	2997	27
OA	911	607	**1001**	**1141**	**1351**	**1491**	1912	2052	2262	2402	2823	2963	10

Bm.n : end of batch.m (time units), n = the number of process line, bold text: the first cycle
Q = queing period (in time units) for each process line in each cycle
t_st_up : start up time   CA : Current Approach
tcyc : cycle time   OA: Optimal Approach

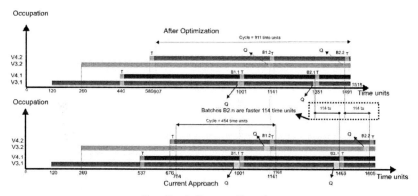

Vp.q: Vessel, Q : queue , T:transfer,
Bm.n : End of batches with m: batch number and n: the line number

Figure 41. Performance evaluation of Plant B.

By converting the accelerated production time to the plant throughput, as shown in Table 23, the throughput increases about 0.00008 batches/time unit or equivalent to the increased productivity of 1.9%. By considering the plant capacity of 10,6 MT/batch, the production time is set in minute and the Biodiesel price of 1,013 USD/MT, the profit of the throughput is expressed in the following:

$$\Delta O(c) = 0.00008 \ x \ 24 \ x \ 60 \ x \ 10.6 \frac{MT}{day}$$

$$\Delta O(c) = 1.221 \ MT/day$$

or equivalent to a value of 1,236.9 USD/day.

Table 23. Throughputs of Plant B.

Path	Throughput [Batches/Time unit]
CA	0.00439
OA	0.00431
Δ Throughput	0.00008 or ↑ 1.9%

Although in terms of productivity, the profit is not as significant, yet there is additional requirement of this plant which cannot be fulfilled by the current operational approach, namely the dangerous idle times which must be minimized by the suitable control strategy. Hence, the following describes in detail the safety specification and how the appropriate control strategy can be figured out.

5.3   Dangerous Idle Times

A generally used approach to ensure quick availability of exclusively used substances in metering tanks is to refill them as soon as possible after allocation. However, in real cases with long time constants, the cycle time of the process could be much longer than its allocation time, so the rough approach generates the idle time of the conflict resources in the metering tanks. Particularly for dangerous resources, the idle times are obviously undesired because of a safety reason. Hence, there is a need to overcome the problem by revising the existing control strategy, but not at all changing the optimal allocation strategy for the shared resources. The revision is realized by an extra procedure to reschedule the refilling process of the dangerous resources. By placing the refilling coinciding as close as possible before the resource allocation, the dangerous idle times can be minimized or even suppressed down to zero. As long as the rescheduling is put into the time interval of the allocation resource cycle which is not influencing the constellation of the fixed optimal allocation strategy, the system remains controlled optimally. Furthermore, the extra procedure to add the safety requirement is described in the following.

At first, the plant is modeled under the initial assumption of the common approach of the quickly refilled resources for all the metering tanks. The plant model is then analyzed to find the optimal sequence. If the state explosion happens, the refinement strategy—as described in the past—may be needed to simplify such a complexity. Hereafter, the obtained optimal control strategy is mapped into an associated Gantt-diagram for further evaluation. The following procedure is then applied.

Step 1: From the obtained optimal Gantt chart, focus on the dangerous resources. By considering the whole time interval, i.e. from the startup up to the cycle, count how long the refilling of each dangerous resource needs to be shifted, so that the refilling precisely coincides with the next resource allocation. The shift times will be used to revise the optimal control strategy as described in the following step.

Step 2: To reduce the behavior of the plant model when the safety specification is added on it, the original model is refined such that it contains only the desired optimal behavior. Next, a signal scheduler module that is used to determine the execution time of the refilling of the dangerous resources is added to the refined model. The signal scheduler model contains the trigger transitions with the input arcs annotated with the shift times of the refilling of the dangerous resources which are obtained from Step 1. By connecting each of the trigger transition to the event input of the transition in the dangerous resource module that serves as "the ignition switch" for the refilling process, the execution of the refilling process can be controlled in accordance to the fixed time parameter. Thus, finally the revised optimal model meets two specifications i.e. the optimal behavior all at once free from the dangerous states.

Step 3: analyze the dynamic graph of the final model to ensure that both desired behaviors (optimal and safe) are already satisfied by the final controller.

A model of a simplified batch-plant as sketched in Figure 42 would be used as an object application for the above procedure. The batch-plant consists of two process lines where each one is connected to a metering tank *MTn* for dosing a kind of dangerous material. In this case, filling (refilling) the material into two metering tanks (MT1 and MT2) is a conflict process and not allowed to be performed simultaneously since they share the same inlet. Each process line allocates the substance in each metering tank for 2 time units and it is then followed as soon as possible by refilling exclusively for 3 time units. Shortly, Figure 43 at the lower part describes the optimal behavior of the original model. By taking a look into the indicators *IDLE 1* and *IDLE 2* which represent the dangerous idle times caused by the common approach of the refilling, the rescheduling on the refilling can be done in the time slots.

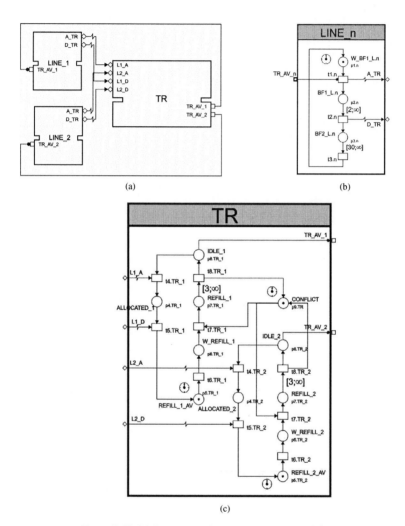

(a)   (b)

(c)

Figure 42. Model of two process lines and the exclusively refilled resource.

Since no idle times occur during the beginning of the process, the rescheduling is not necessary at the initial time. The idle time arises after the first allocation until then the process enters its steady state condition. By considering that the optimal cycle time is 32 time units and 5 time units are totally required for the allocation time as well as the refilling duration, accordingly the refilling can be shifted as far of 27 time units relatively from the previous allocation time, as shown by the targeted Gantt chart at the upper part of Figure 43.

Additionally, to reduce the behavior of the original model when the rescheduling is modeled, the refinement of the original model is necessary. The refined model is further revised by adding the module of the signal scheduler for the refilling of the dangerous resource.

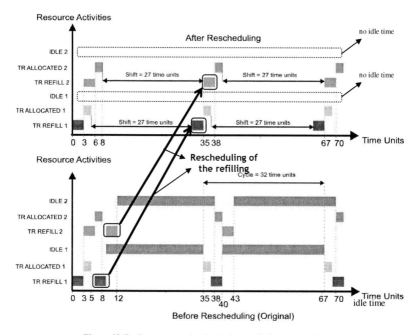

Figure 43. Performance evaluation before and after the rescheduling.

Figure 44 describes the refined model which is revised by the module of the signal scheduler named by *TR_SC* (see Figure 44(c)). In the module, the place *W_FILL_L1* annotated without the time delay represents zero waiting time during the initial filling of the first metering tank, whereas the arc of the place *W_FILL_L2* timed with 3 time units represents the delay of the filling of the second metering tank (because of the conflict with the filling of the first metering tank). When the resources are allocated by the main processes, the current tokens stay in the places *ALOC_L1* or *ALLOC_L2* until receiving the dealocating signals from the main modules. The tokens then move onto the next state i.e. places *W_REFILL_L1* and *W_REFILL_L2* for 27 time units, waiting for the upcoming refilling. Finally the resources are reallocated following the cycle of the main process lines.

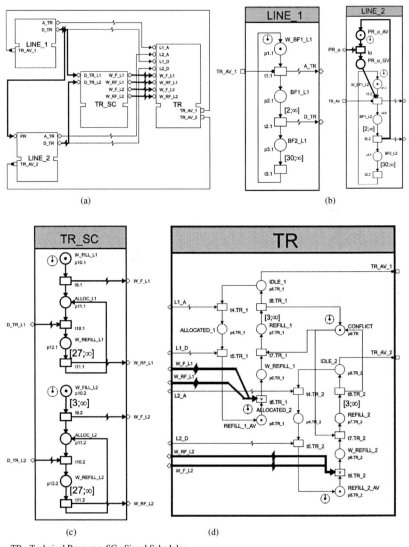

(a)   (b)

(c)   (d)

TR : Technical Resource, SC : Signal Scheduler

Figure 44. The revised refined model for minimizing the dangerous idle times.

By the last step, the revised optimal model is analyzed. Comparison among the process activities before and after the rescheduling indicates that the dangerous idle times are successfully forced to be zero, as expected by the indicators *IDLE 1* and *IDLE 2* shown in Figure 43 at the upper part. The cycle time of the revised schedule keeps the same with the original one i.e. 32 time units, which means that the optimal behavior of the system is unchanging at all after the rescheduling.

Dangerous Idle Times on Plant B

Because the problem of the dangerous idle times exhibits only in the second plant (Plant B), the rescheduling is only focused on this plant. As already described before, the obtained optimal sequence uses the assumption of the quickly refilled resources. The optimal model of the plant is revised by including the signal scheduler module which contains the shift times of the dangerous resource refilling (substance B—modeled by the resource module TR4) as marked by S in Figure 46. The signal scheduler module is connected to the resource module which can be seen partially in Figure 45, while its complete version can be seen in Appendix A2.3. It is obvious, that the refilling activities are set by depending on the token flow in the module *TR4_SC* which is parameterized by the shift times (S). As for example, the initial filling of the first resource *MT1* shifted for 95 time units (as indicated by S in Figure 46) is denoted by the first arc timed with the number of 95. Likewise with the rescheduling for the following refilling activities, they can be traced easily from the model, i.e. for $S = 295$ time units, $S = 516$ time units and $S = 305$ time units. Consider also the shifts of the refilling of the second metering tanks, namely $S = 235$ time units, $S = 295$ time units, $S = 516$ time units and $S = 305$ time units.

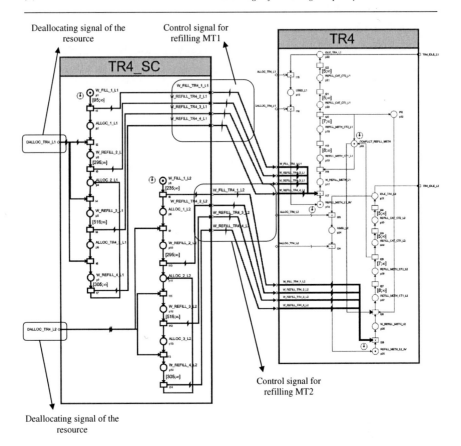

Figure 45. Model of the signal scheduler for refilling the dangerous resource TR4.

The performance evaluation on the revised optimal model indicates that the desired behavior of the plant, as expected by the Gantt-diagram in the upper part of Figure 46, can be achieved. The idle times *TR4 IDLE L1* as well as *TR4 IDLE L2* are successfully reduced down to zero without affecting the optimal cycle time of the plant production. By cross-checking the end of each batch (Bm.n) before and after replacement of the resource refilling TR4, it is obvious that the optimal production time is not changing at all after the rescheduling.

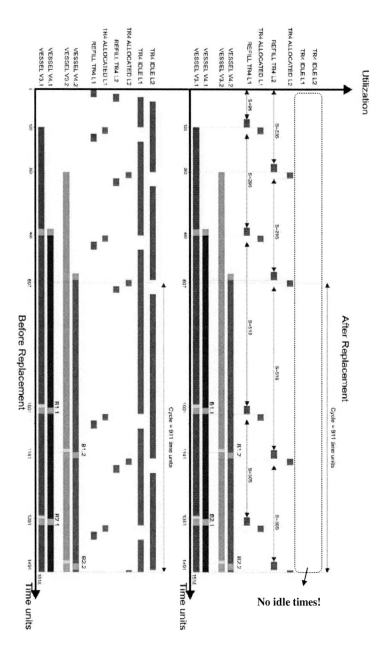

S=Shift (in time units),TR: Technical Resource, Bm.n : End of batches with m: batch number and n = line number

Figure 46. Performance evaluation of Plant B after rescheduling the refilling process of the dangerous resource TR4.

5.4   Globally Optimal Behavior

The final challenge of this thesis is to unify two batch-plants coupled by one shared resource into one integrated scheduling system where the task is not relatively simple. The unification problem arises from the use of the conflict resource, namely the availability of the pipeline allocation for transferring Biodiesel from each plant to the final storage tank, which can alter the optimal production behavior when both plants are unified. However, by using a buffer storage tank (the daily tank) installed in Plant B, the products of plant B can be stored into the temporary storage (waiting), meanwhile at the same time the allocation of the pipeline can be given for the products of Plant A. The goal of the final controller is to determine when the waiting times in the buffer tank are required, so that it leaves the controller of each plant running independently and optimally.

To ensure that the dynamic graph of the both plants only contains the optimal behaviors, there is a need to model a restriction for the optimal control strategy. Although the concept is not new, special transitions so-called *facts* have been developed in timed arc Petri nets [8,9], yet in the context of application of the used modeling method as well as this case, the idea of the control restriction gives a new alternative approach. The restriction is realized by forbidden transitions which will detect the change of the optimal behavior of the modeled systems. The key of preventing the forbidden behavior is the usage of inhibitor arcs which will disable the connected transitions when having a true value, so as once a forbidden transition fires, no following states are computed in the dynamic graph. Finally the dynamic graph becomes simpler and only contains the permissive behavior (optimal). Clearly, the following describes the corresponding model.

Consider a simplified model as sketched in Figure 47 which is purely transformed from an optimal resource allocation strategy of a process example. The optimally waiting times are assumed to be already known and denoted with $\tau1,...,\tau m$ attached at the waiting places $W1,..., Wm$, while the ongoing sequence of the resource allocation are represented by the places $ALLOC\_1 ... ALLOC\_m$. Once the needed resource is available (as reported by condition input $RESOUERCE\_AV$), the request of allocation is realized by the event signals $A1\_TR... Am\_TR$ sent to the resource module (omitted from the model). After the resource is used, the event signals $D1\_TR... Dm\_TR$ release the deallocating signals to stop the allocation activity. In case of violation of the control restriction, i.e. when the resource is unavailable on the requested time (because of being used by another process unit), so a forbidden transition $(FORB\_Tm)$ fires followed soon by releasing the corresponding event signal (DS_m) to the module *Dead State*. As a consequence, the current state *ALIVE* changes to the next state

DEAD which forces a true value at the condition input *DEAD_T* to activate all the connected inhibitor arcs. Then, the system is deadlock and no following states are computed in the dynamic graph.

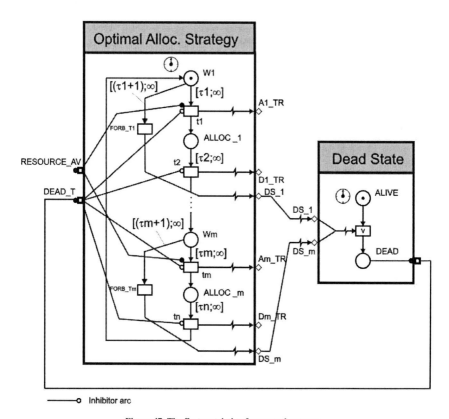

Figure 47. The first restriction for control strategy.

The simplified model of the unified plants by including the control restriction can be seen in Figure 48. As expressed by the model, the module *Optimal Alloc. Strategy (A)* represents the fixed optimal behavior of plant A while module *Optimal Alloc. Strategy (B)* denotes the fixed optimal resource behavior of plant B. To prevent conflicts when both plants simultaneously access the final storage tank—automatically generating the forbidden states, then in the module *Buffer Tank,* a waiting mechanism expressed by the place *WAITING* is added. The waiting place indicates that the product transfer of Plant B, denoted by the place *TRANS_TO_FS,* must wait temporarily until the product transfer of Plant A is fully accomplished and the pipe line is ready to be accessed. With the existence of the waiting mechanism (the buffer tank), the optimal production cycle of both plants remains unaffected at all.

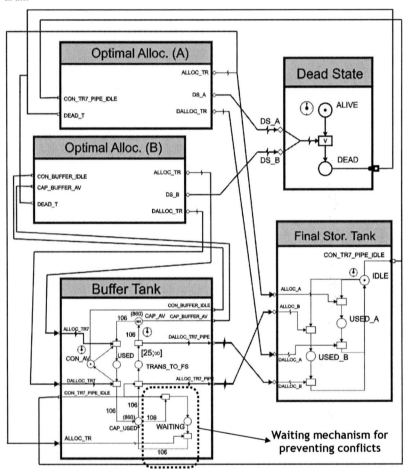

Figure 48. Simplified model of the unified plants.

Nevertheless, when trying to analyze the cyclic model of the unified plants, the resulting states are still big i.e. 66355 states such that the dynamic behavior is difficult to be analyzed. This is actually understandable since the computed dynamic graph must deal with the fixed different cycle times as well as the fixed different startup times of the both plants such that the global cycle is hard to be created. Therefore, the limited cyclic model (the cycle behavior being finite with respect time), as has been introduced in Figure 27, must be used. However, the restriction will not be changing the optimal plant behavior and is exactly appropriate with reality on the factory where the plants are actually operated based on the market demand limited. Clearly, the second restriction for the allocation strategy is modeled in Figure 49, as a result of modifying the model in Figure 47. By setting the input of the operation requests (denoted by the place *REQ*) flexibly, for instance: 10 batches, 20 batches, 100 batches and so on, the cyclic behavior is limited. The place *REQ* will release one by one token into its succeeding place, i.e. *REQ_A,* until all requests of the allocation are accomplished. Finally, the transition *REQ_END* removes the main token from the place *W1,* and the cycle of the process is terminated. By this approach, the unified plant behavior is easily analyzed. As for example, when the input of the limited cyclic model is set for 100 batches, the resulting dynamic graph of both plants contains only 1922 states as shown in Figure 50. Compare with 66355 states of the unlimited cyclic model.

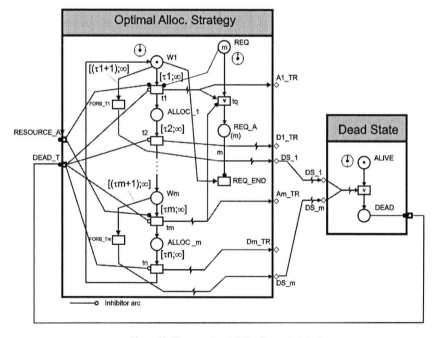

Figure 49. The second restriction for control strategy.

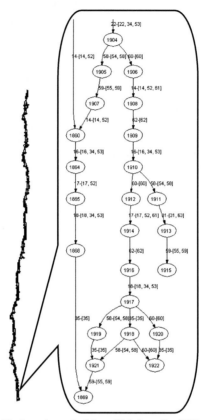

Figure 50. The dynamic graph of the unified plants (for 100 Batches).

Figure 51 represents the optimal Gantt-diagram for ten batches of the unified plants. By comparing the result with the data of the optimal production time of each plant prior to the unification, as given in Table 24, it is obvious that the optimal behavior of each plant does not change at all after both plants are unified. In other words, the plants are successfully optimized by the proposed control strategy as the main goal of this project.

Table 24. Data of the optimal production time of each plant prior to the unification (in time units).

Products	B1.1	B1.2	B2.1	B2.2	B3.1	B3.2	B4.1	B4.2	B5.1	B5.2
Plant A	984	1086	1457	1559	2001	2103	2489	2592	3033	3136

Products	B1.1	B1.2	B2.1	B2.2	B3.1	B3.2	B4.1	B4.2	B5.1	B5.2
Plant B	1001	1141	1351	1491	1912	2052	2262	2402	2823	2963

Bm.n : Product of each Plant
m: end of each batch
n : the line number of each plant

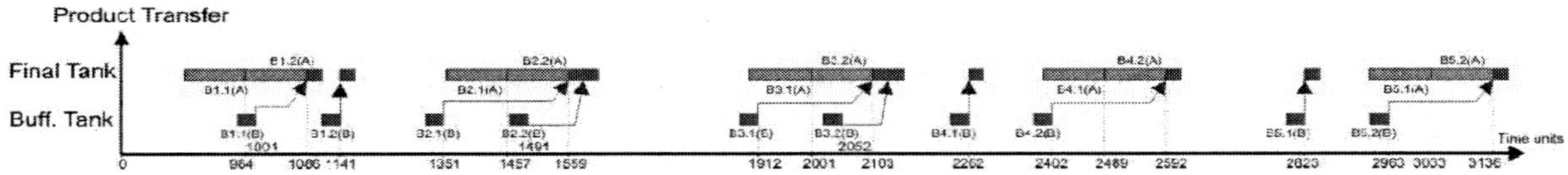

Figure 51. Utilization of the buffer tank and the final storage tank regarding the product transfers after the both plants are unified.

Bm.n(B): Product of Plant B(Keep optimal after the both plants are unified)
Bm.n(A): Product of Plant A (Keep optimal after the both plants are unified)

m: end of each batch, n : the line number

6.   Conclusions

This work has demonstrated an application of the methods that have been developed by the author for an industrial-sized example. There are some novel results of the proposed approaches.

The first is the fact that the approaches are successfully applied to real manufacturing plants in process industry. This means that one does not deal with some kind of "artificially" constructed academic example, but with real plants of real scale and complexity.

The second is to provide a realistic model of large plants by using systematic and structured ways as recommended by the batch standards. By such an approach, the modeling complexity can be simplified and one obtains results in modular forms that are strongly sufficient with real applications in the industrial world. Beside of that, the timing concept existing in the used modeling method enables us to develop models that can behave in accordance to actual timing specifications of real plants, thus performance of the plants can be simulated precisely.

Last but not least, the plant specifications, corresponding to process optimization and a safe operation as the primary goals of the presented case, must be defined formally. The formalization covers two criteria, i.e. waiting periods of the used process devices as well as dangerous idle times. By the formally defined criteria, the controller can be analyzed in accordance to the manufacturing requirements.

Controller analysis for real plants must deal with various complexity problems. As inducted from the investigated case, the complexity problem starts with determination of an optimal sequence of each single production line of each plant, which faces the complicated paths, as a consequence of exponentially growing nodes following the existing control recipe options. By using the closest occupation times among the consecutively used process devices, the complexity can be simplified. Likewise with the state explosion problem following the coupling of the single line schedules into the common scheduling system of each plant, the refined original model can handle such issue. Particularly for dangerous idle times that must be minimized, by revising the existing control strategy, i.e. by rescheduling the refilling of the dangerous resource coinciding as close as possible prior to the resource allocation (yet not changing the optimal resource allocation), the dangerous idle times can be successfully minimized. Finally, to simplify the complexity of unifying the both plant schedules, restrictions for the optimal control strategy are proposed. There are two restrictions applied. The first restriction is realized by forbidden transitions which will detect the change of the optimal behavior of the modeled systems. When a forbidden transition fires, no following states are computed in the dynamic graph such that at the end the dynamic graph only contains the permissive behavior (optimal). The second restriction is the use of the limited cyclic model, where the cycle behavior is assumed to be finite with respect to time. By these restrictions, the dynamic behavior of the unified plants becomes simpler and easier to be analyzed thus, the desired control strategy can be synthesized. As the main contributions, the proposed methods can tackle the problems and these are proven to be able to achieve the expected results.

The achieved results indicate improvements. The resulting dynamic graphs can be reduced significantly and even decreasing the states more than 90%. The performance analysis on both plants shows that the dangerous idle times can be suppressed down to zero. The increase of the plant throughputs, approximately 9550.9 USD a day, can be achieved by unifying both plants into the globally scheduling management without changing the optimal behavior of the production systems at all. Last but not least, the modeling and analysis of both plants gives transparent information of process sequences such that it can be used to improve the manually controlled systems existing in the plants.

References

[1]   D. Azzopardi, S. Lloyd, "Scheduling and Simulation of Batch Process Plant through Petri net modeling", Factory 2000: 4th IEEE International Conference on Advanced Factory Automation, York, Oct. 3-5, 1994, Proceedings.

[2]   S.M. Clark, G.S. Joglekar, "Features of Discrete Event Simulation, in Batch Processing Systems Engineering, Fundamentals and Applications for Chemical Engineering",Edited by G. V. Reklaitis, A. K. Sunol, D. W. T. Rippin and Ö. Hortacsu,NATO ASI Series, Series F: Computer and Systems Sciences, Vol. 143, Springer Verlag, 1996.

[3]   M. Fritz, K. Preuß, S. Engell, "A Framework for Flexible Simulation of Batch Plants", Proceedings of the 3rd International Conference on Automation of Mixed Processes, ADPM 1998, Reims, France, 19.-20.3.1998.

[4]   A. Liefeldt, T. Löhl, M. Stobbe, S, Engell, "Simulation and Scheduling of Recipe-Driven Batch Processes based on a Single Data Model", Interkama ISAtech Conference, Düsseldorf, Germany, October 18-20, 1999.

[5]   H.M. Hanisch, "On the Use of Petri Nets for Design, Verification and Optimization of Control Procedures for Batch Processes", IEEE Conference on Systems, Man and Cybernetics, San Antonio, Texas, October 1994, Proceedings, Vol. 1, pp.326-330.

[6]   M.-F. Amir, H.-M. Hanisch, "Modeling and Simulation of Optimal Scheduling of a Biodiesel Batch-Plant: Modular Strategy for Tackling Complexity", International Conference Instrumentation Control and Automation ICA 2009, Bandung-Indonesia, October 2009.
       Available on: http://ica-itb.org/2009/archive/ICA2009-A12.pdf

[7]   Ulrich Christmann, "BatchMon - Monitoring and Simulation of Recipe Driven Batch Processes in Disturbance Situations", Ph.D thesis, ISBN 3-89722-801-7, 2001.

[8]   H.-M. Hanisch, "Analysis of Place/Transition Nets with Timed Arcs and its Application to Batch Process Control", Lecture Notes in Computer Science, Vol. 691, pp. 282-299, Springer-Verlag, 1993.

[9]   H.-M. Hanisch und U. Christmann, "Modeling and Analysis of a Polymer Production Plant by Means of Arc-Timed Petri Nets", Conference on Computer Integrated Manufacturing in the Process Industries (CIMPRO '94), Rutgers University, New Brunswick, NJ, April 1994, Proceedings, 194-207, also in : International Journal of Flexible Automation and Integrated Manufacturing, 3(1), pp. 33-46, 1995.

[10]   Tianlong Gu, Parisa A. Bahri, "A survey of Petri net applications in batch processes", Comput. Ind. 47 (2002), pp. 99–111.

[11]   M. Fritz, K. Preuß, S. Engell, "A Framework for Flexible Simulation of Batch Plants", Proceedings of the 3rd International Conference on Automation of Mixed Processes, ADPM 1998, Reims, France, 19.-20.3.1998.

[12]   M. Fritz, A. Liefeldt, S. Engell, "Recipe-Driven Batch Processes: Event Handling in Hybrid System Simulation", ISATech Conference, Düsseldorf, Germany, October 18-20, 1999.

[13]   R.R.H. Schiffelers, D.A. van Beek, J. Meuldijk, J.E. Rooda, "Hybrid modelling and simulation of pipeless batch plants", Editors, *Computer-Aided Chemical Engineering* (2002).

[14]   Tianlong Gu, Parisa A.Bahri, Guoyong Cai, "Timed Petri Net Based Formulation and an Algorithm for the Optimal Scheduling of Batch Plants", Int.Journal Appl. Math. Comp. Sci. ,2003, Vol. 13 No 4, 527-536.

[15]   H. Brettschneider, H.J. Genrich, H.M. Hanisch, "Verification and Performance Analysis of Recipe-based Controllers by means of Dynamic Plant Models", 2nd International Conference on Computer Integrated Manufacturing in Process Industries (CIMPRO '96), Eindhoven, The Netherlands, June 3-4, 1996, Proceedings, pp. 128-142.

[16]   H.J. Genrich, H.M. Hanisch, K. Wöllhaf, "Verification of Recipe-Based Control Procedures by Means of Predicate/Transition Nets", Lecture Notes in Computer Science, Vol. 815, Springer Verlag, 1994, pp. 278-297.

[17]   NE33 – Requirements to be met by systems for recipe-based operations. Namur WG 2.3: Functions of Operations Management and Production Control Level,1992.

[18]   S88 Batch Control Part 1: Models and Terminology, ISA-S88.01-1995, ISA, Standard, February 1995.

[19]   H.M. Hanisch, S. Fleck, "A Resource Allocation Scheme for Flexible Batch Plants based on High-Level Petri Nets", IEEE SMC, CESA `96 IMACS Multiconference, Lille, France, July 9-12, 1996.

[20]   Martin Mittelbach and Claudia Remschmidt," Biodiesel – The Comprehensive Handbook" ISBN: 3-200-00249-2, 2004.

[21]   H.-M. Hanisch: Petri-Netze in der Verfahrenstechnik, Verlag R. Oldenburg, München, Wien, 1992.

[22]   T. Murata, "Petri Nets: Properties, Analysis and Applications", Proceedings of the IEEE, Vol. 77, No. 4, April 1998, pp. 541-580.

[23]   R. David, "Modeling of Dynamic Systems by Petri Nets", Proceedings of the ECC 91 European Control Conference, Grenoble, France, July 2-5, 1991, pp. 136-14.

[24]   R. David, H. Alla, "Petri Nets and Grafcet: Tools for Modelling Discrete Event Systems", Prentice Hall, London, 1992.

[25]   J.M. Proth, X.L. Xie, "Petri Nets: A tool for design and management of manufacturing systems", John Wiley & Sons, 1996.

[26]   R.S. Sreenivas, B.H. Krogh, "On Condition/Event Systems with Discrete State Realisations", Discrete Event Dynamic Systems: Theory and Applications, 2 (1), 1991, pp. 209-236.

[27]   J. Thieme, "Symbolische Erreichbarkeitanalyse und automatishce Implementierung structucturierter zeitbewerteter Steuerungsmodelle", Hallenser Schriften zur Automatisierungstechnik, Bd.3, Logos-Verlag, Berlin, Dissertation, 2002.

[28]   H.M. Hanisch, J. Thieme, A. Lüder, O. Wienhold, "Modelling of PLC Behavior by Means of Timed Net Condition/Event Systems", 6th IEEE International Conference on Emerging Technologies and Factory Automation ETFA `97, Los Angeles,USA, September 1997.

[29]   S. Kowalewski, J. Preußig, "Timed Condition/Event systems, "a framework for modular discrete models of chemical plants and verification of their real-time discrete control", 2nd International Workshop on Tools and Algorithms for the Contruction and Analysis of Systems (TACAS `96), Passau, Germany, March 27-29, 1996.

[30]   H.-M. Hanisch, J. Thieme, A. Lüder, "Toward a Synthesis Method for Distributed Safety Controller Based on Net Condition/Event Systems", J. Intell. Manufact., Vol. 8, No.5, pp. 357-368, 1997.

[31]   H.-M. Hanisch, A. Lüder, and M.Rausch, "Controller Synthesis for Net Condition/Event Systems with a Solution for Incomplete State Observation", Eur. J.Contr., Vol. 3,pp. 280-291,1997.

[32]   V. Vyatkin and H.-M. Hanisch, "A Modeling Approach for Verification of IEC 1499 Function Blocks Using Net Condition/Event Systems", in Proc. 7th IEEE 37 th Conf. Decision Control, Tampa, FL, Dec. 16-18, 1998, pp. 3218-3223.

[33]   U. Christmann, H.-M. Hanisch, "BatchMon – Applying TNCES-Models to the Online Monitoring of of Recipe Driven Batch Plants", invited to: IEEE Conferences on Systems, Man and Cybernetics (SMC '98), San Diego, California, Oct. 11-14, 1998.

[34]   Missal, D. and Hanisch, H.-M., "A modular synthesis approach for distributed safety controllers, part a: Modelling and specification," In: 17th IFAC World Congress, proceedings. Seoul, Korea, IFAC, July 2008, pp. 14 473–14 478.

[35]   Missal, D. and Hanisch, H.-M., "A modular synthesis approach for distributed safety controllers, part b: Modular control synthesis," In: 17th IFAC World Congress, proceedings, Seoul, Korea, July 2008, pp. 14 479–14 484.

[36]   Missal, D. and Hanisch, H.-M., "Modular Plant Modelling for Distributed Control". In: Proceedings of 2007 IEEE International Conference on Systems, Man and Cybernetics, Montréal, Canada, October 2007, pp. 3475-3480.

[37]   H.-M. Hanisch, M. Hirsch, D. Missal, S. Preuße, C. Gerber, " One Decade of IEC 61499 Modeling and Verification - Results and Open Issues", Preprints of the 13th IFAC Symposium on Information Control Problems in Manufacturing, Moscow, Russia, June 3 - 5, 2009.

[38]   Christian Gerber, H.-M. Hanisch, "Timed-Net-Condition-Event-System Workbench, Documentation of the functionality and the implementation at SWI-Prolog", Institut Informatik Lehrstuhl fuer Automatisierungstechnik, MLU Halle-Wittenberg, 2007 (unpublished).

[39]   The data of benchmark prices of Biodiesel export sourced from Ministry of Industry and Trade the Republic of Indonesia, available at: http://www.depdag.go.id/files/regulasi/2010/10/HPE27102010.pdf

[40]   Christian Gerber, " Implementation and Verification of Distributed Control Systems", Hallenser Schriften zur Automatisierungstechnik, Bd.7 ,Logos-Verlag, Berlin, Dissertation, 2011 (will be published).

[41]   O. Gutzeit, "Modellbasierte Entscheidungsunterstützung bei der Fertigung bahngeführter Materialien", Hallenser Schriften zur Automatisierungstechnik, Bd.5 , Logos-Verlag, Berlin, Dissertation, 2009.

[42]   H.-M. Hanisch, A. Lüder, " A Signal Extension for Petri nets and its Use in Controller Design", In: Burkhard, H.-D.; Czaja, L.; Starke, P.: Informatik-Berichte, No. 110: Workshop Concurrency, Specification and Programming, 28-30 September 1998, pages 98-105. Berlin: Humboldt-Universität, 1998.

[43]   H.-M. Hanisch, A. Lüder, "Modular Modeling of Closed-Loop Systems", In: Proceedings of Colloquium on Petri Net Technologies for Modelling Communication Based Systems, October 21-22, 1999, pages 103-126. Fraunhofer Gesellschaft, ISST, 1999.

Appendix

Appendix A1. Closed-Loop Model of Both Plants

A1.1 Model of each single production line of Plant A

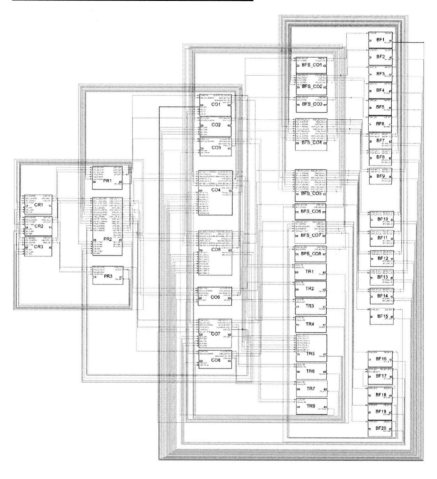

A1.2 Model of Plant A (two single production lines)

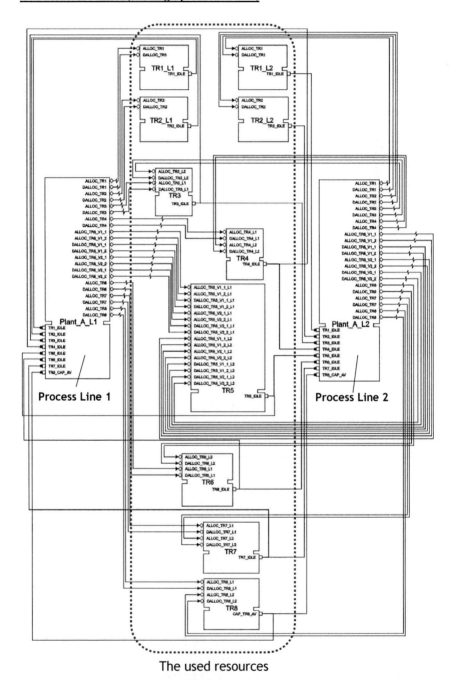

The used resources

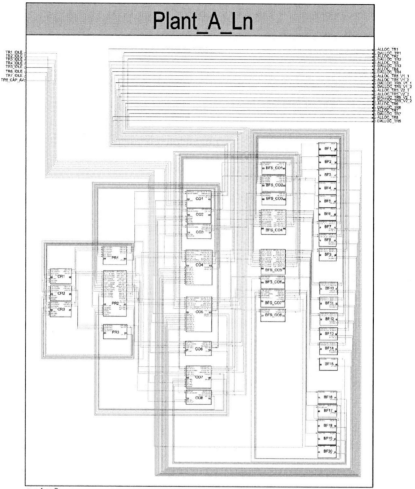

n = 1 or 2

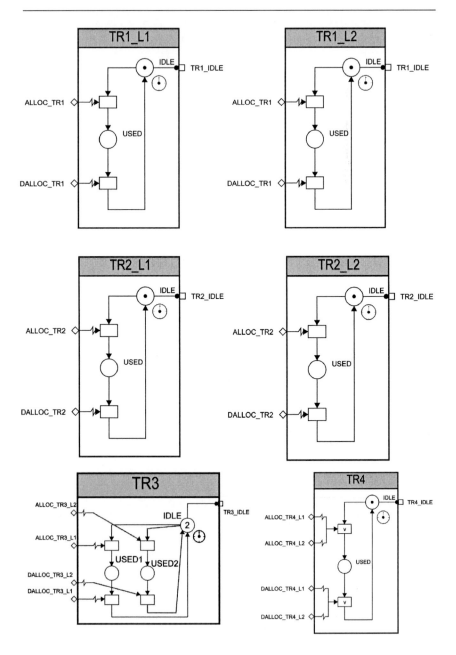

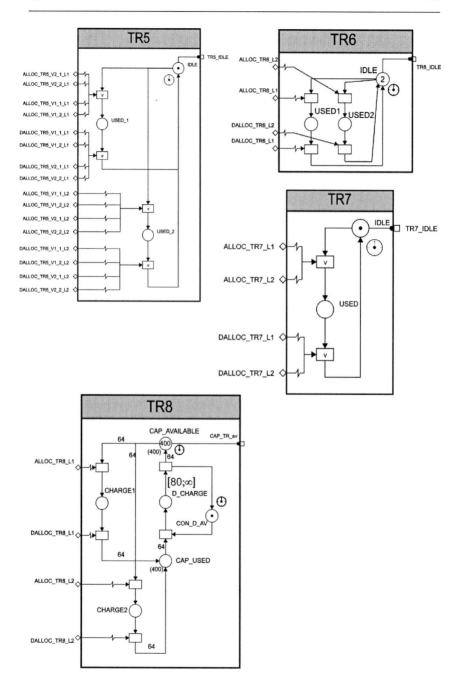

A1.3 Model of each single production line of Plant B

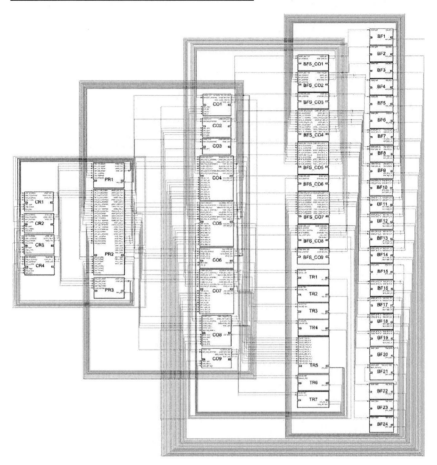

A1.4 Model of Plant B (two single production lines)

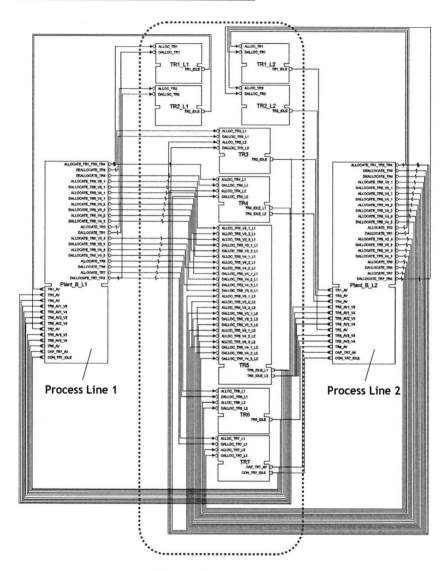

The used resources

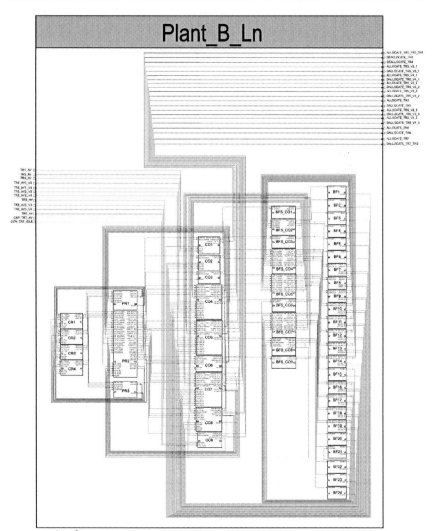

n = 1 or 2.

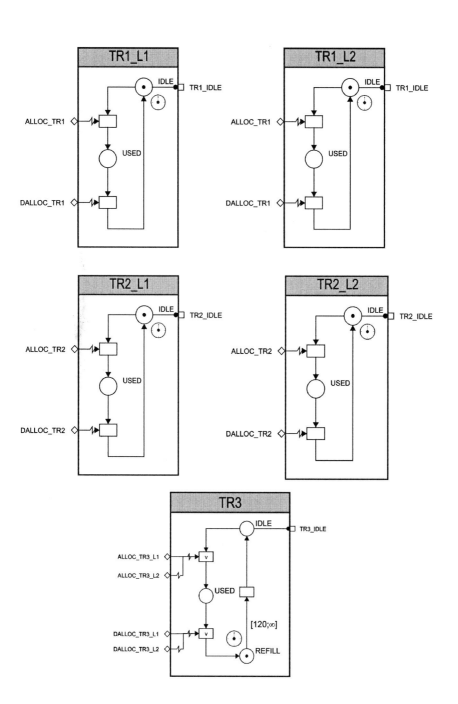

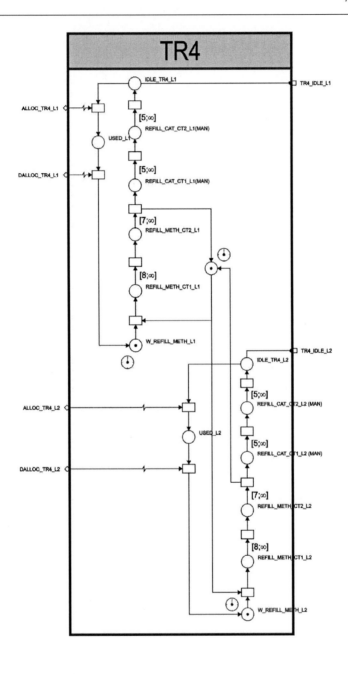

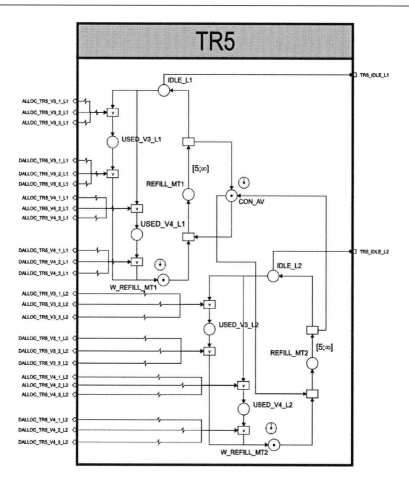

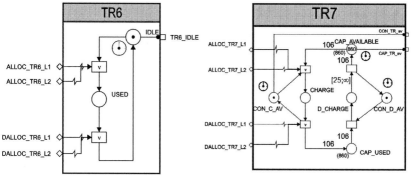

Appendix A2. The Refined Model of Both Plants

A2.1 The Refined Model of Plant A

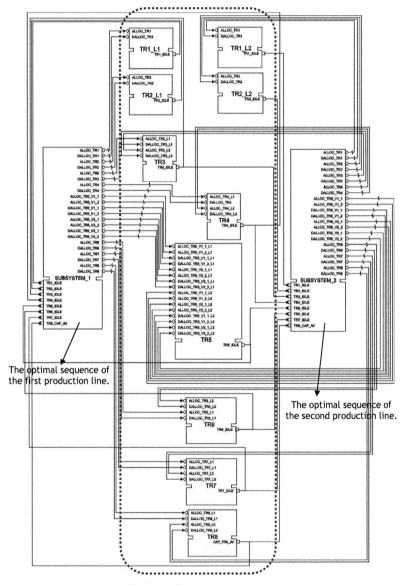

The used resources, see
Appendix A1.2.

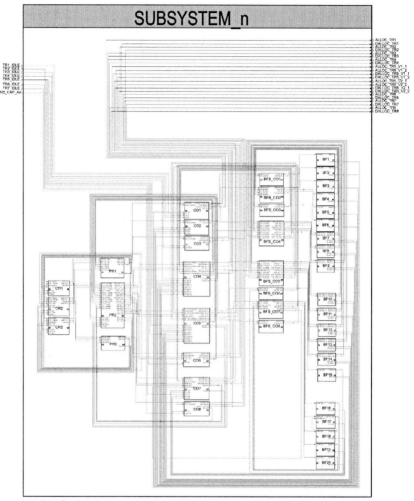

n = 1 or 2

A2.2 The Refined Model of Plant B

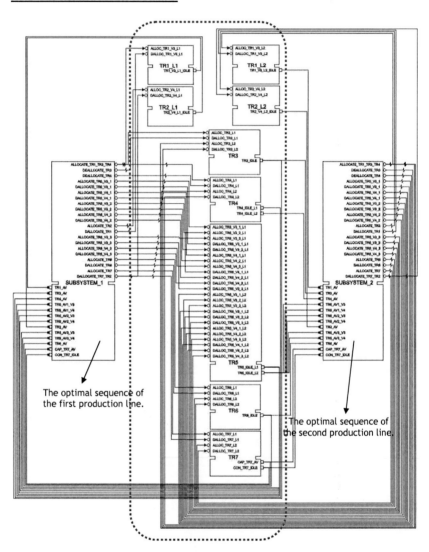

The optimal sequence of
the first production line.

The optimal sequence of
the second production line.

The used resources, see
Appendix A1.4.

SUBSYSTEM_n

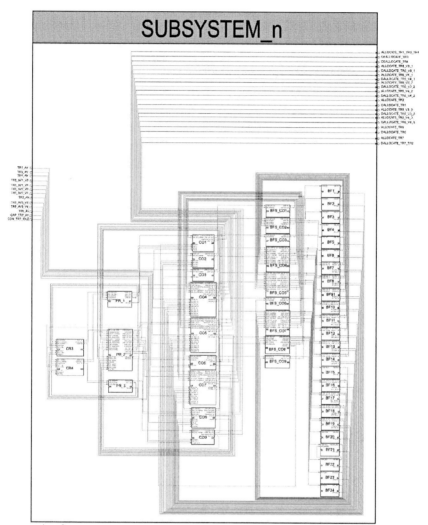

n: 1 or 2.

A2.3 The Refined Model of Plant B with the Signal Scheduler

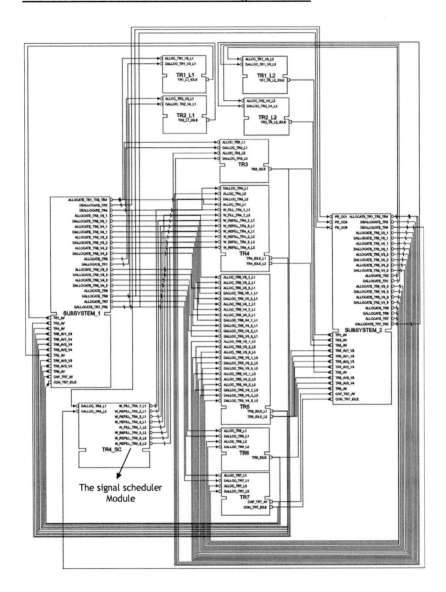

The signal scheduler
Module

Biographical Sketch

Mohamad Fauzan Amir
Born in 1977 in Kediri (Indonesia)

Education

1984-1990	Elementary School (in Indonesia)
1990-1993	Secondary School (in Indonesia)
1993-1996	High School (in Indonesia)

Study

1996-2002	Study in Bachelor Program of Engineering Physics
	Institut Teknologi Bandung
	Indonesia
2002	Degree: B. Eng. (Bachelor of Engineering)
2003- 2005	Study in Master Program of Instrumentation and Control
	Institut Teknologi Bandung
	Indonesia
2005	Degree: M. Eng. (Master of Engineering)

Professional Career

2002-2007	Teaching Assistant/Training Staff
	Department of Engineering Physics
	Institut Teknologi Bandung
	Indonesia
2007 - 2011	Scientific Staff (under the support of the DAAD)
	Chair for Automation Technology
	Martin Luther University of Halle-Wittenberg
	Germany

Selbstständigkeitserklärung

Hiermit erkläre ich, dass ich diese Arbeit selbstständig und ohne fremde Hilfe verfasst habe. Es wurden keine anderen als die von mir angegebenen Quellen und Hilfsmittel benutzt. Die den benutzten Werken wörtlich oder inhaltlich entnommenen Stellen sind als solche kenntlich gemacht.

Die Arbeit wurde bisher weder im In- noch im Ausland in gleicher oder ähnlicher Form einer anderen Prüfungsbehörde vorgelegt.

Halle (Saale), den 24. Mai 2011

Mohamad Fauzan Amir

Abstract

Generally, scheduling problems accompanying typical batch processes are vitally important to be solved for improving the plant productivity. In these respects, finding a good and feasible schedule or even an optimal result, by which costs and lead times can be reduced, is often a very complex and also a difficult task. Moreover, in large plants, the challenges come not only from the modeling ways that require systematic and structured approaches, but also from the exact strategies how the performance of the model can be analyzed. The goal of this research is to develop a comprehensive study on industrial-sized plants, with regard modeling and analysis of scheduling problems. Formalization of the required plant specifications, the modularly modeling ways—which refer to the widely used batch standards, and also the strategies for tackling complexity, are the main contributions of this thesis. These studies will be carried out by using the Timed Net Condition/Event Systems (TNCES) model. Finally, the model is analyzed to synthesize an optimal control strategy for the investigated plants.